PRIONS
The New Biology
of Proteins

PRIONS
The New Biology of Proteins

CLAUDIO SOTO

CRC Press
Taylor & Francis Group
Boca Raton London New York

CRC Press is an imprint of the
Taylor & Francis Group, an **informa** business

A TAYLOR & FRANCIS BOOK

Cover illustration drawn by Eric D. Smith, Harvard University.

CRC Press
Taylor & Francis Group
6000 Broken Sound Parkway NW, Suite 300
Boca Raton, FL 33487-2742

First issued in paperback 2019

© 2006 by Taylor & Francis Group, LLC
CRC Press is an imprint of Taylor & Francis Group, an Informa business

No claim to original U.S. Government works

ISBN-13: 978-0-8493-1442-1 (hbk)
ISBN-13: 978-0-367-39145-4 (pbk)

Library of Congress Cataloging-in-Publication Data

Catalog record is available from the Library of Congress

Visit the Taylor & Francis Web site at
http://www.taylorandfrancis.com

and the CRC Press Web site at
http://www.crcpress.com

Dedication

To Claudia, Andres, Kevin, Dylan, and my little girl to be....

Preface

Transmissible spongiform encephalopathies (TSEs), also known as prion-related diseases, are a group of infectious, fatal neurodegenerative disorders for which there is no cure, treatment, or early diagnosis. TSEs are dramatic diseases that rapidly, progressively, and inexorably destroy the cognitive, motor, and sensorial skills that are the essence of human beings. At the molecular level, the disease is likely caused by the misfolding of the prion protein, which accumulates in the brain and produces neurodegeneration. Several unprecedented scientific findings, which have directly confronted popular dogmas in biology, have put prion research in the spotlight. The experimental evidence strongly supports an entirely novel disease mechanism, involving disease transmission by replication of the misfolding of a single protein in the absence of nucleic acids. The popularity of prion diseases is also due to the panic produced by the recent appearance of a new human disease (variant Creutzfeldt-Jakob disease) that is transmitted by eating meat contaminated by bovine spongiform encephalopathy (BSE), better known as mad cow disease. Because of insufficient information available regarding the incubation time and the actual level of exposure to the contaminated material, it is impossible to make any well-founded prediction about the future of this nascent epidemic.

This book attempts to combine a detailed and up-to-date description of the state of the knowledge in the field with an intriguing but tempered speculation of the putative implications of the findings to our current understanding of biology. During the last few years we have begun to perceive the broader implications of the heretically attractive prion concept of transmission of biological information by propagation of alternative protein folding. I hope that this book can contribute to liberating the imagination of the reader to see the new scientific world opened by prions, which impact broader areas of biology, public health, and biotechnological strategies for therapy and diagnosis.

Although I regretted it many times during the long process of preparing this book, I decided to write it entirely myself in order to give continuity to the text, maintain a homogenous style, and provide an appropriate connection between chapters. The decision was also guided by a desire to avoid important gaps and repetitions, which are frequent in multiauthored books. I could not have completed this book without the constant support of my family, lab members, and close collaborators. To them I extend my sincere gratitude for their patience, help, and illuminating discussions.

Author

Claudio Soto is currently the Charlotte-Warmoth professor of neurology, human biological chemistry and genetics, and neuroscience and cell biology, and director of the Protein Misfolding Disorders Research Unit at the University of Texas Medical Branch in Galveston. Dr. Soto was recently awarded the Green Distinguished University Chair in Neurosciences and was appointed Director of the new George and Cynthia Mitchell Center for Alzheimer's Disease Research. He received his Ph.D. in biochemistry and molecular biology from the University of Chile in 1992 and was a postdoctoral fellow at the Catholic University of Chile and at the New York University School of Medicine, where he became an assistant professor of research in 1995. Between 1999 and 2003, Dr. Soto was senior scientist, head of the Department of Molecular Neuropathology, and scientific advisor for neurobiology at Serono International in Switzerland. For the past 12 years, he and his colleagues have engaged in research into the molecular basis of neurodegenerative diseases associated with the misfolding of proteins and their accumulation in the brain, focusing particularly on Alzheimer's and prion-related disorders. His work has led to the development of novel strategies for treatment and diagnosis of neurodegenerative diseases.

Dr. Soto has published more than 80 scientific articles in some of the most prestigious peer-reviewed journals. He has received numerous awards and has been invited to give presentations at more than 50 international meetings worldwide. Dr. Soto is a member of the editorial board of several scientific journals and has acted as adviser and reviewer for many institutions, including the World Health Organization, the U.S. National Institutes of Health, and the European Commission, among others.

Contents

chapter one

Human and animal diseases: clinical symptoms, epidemiology, and neuropathology

Prion diseases are a closely related group of fatal neurodegenerative disorders that affect humans and other mammals [Collinge, 2001]. They have also been called transmissible spongiform encephalopathies (TSEs), slow virus diseases, and subacute spongiform encephalopathies. A unique feature of these diseases is that they can have three different origins: sporadic, inherited, and infectious. The clinical, epidemiological, and neuropathological features can be very different in each of the diseases, but they are classified together because the key molecular event appears to be the same, i.e., the misfolding of the prion protein [Collinge, 2001; Prusiner, 1998]. This chapter provides an overview of the main characteristics of different forms of prion diseases, describing their origin, clinical symptoms, prevalence, and neurological alterations.

1.1 Human diseases

Creutzfeldt-Jakob disease (CJD) is the most common form of TSE in humans. The first description of the disorder is attributed to H.G. Creutzfeldt (1920), although by current diagnostic criteria the case reported by Dr. Creutzfeldt would not classify as CJD. In 1921, A. Jakob described four similar cases, and at least two of them had clinical features suggestive of CJD. Since then, CJD cases have been recorded all over the world, and it is now recognized that the disease has a worldwide distribution [Johnson and Gibbs, Jr., 1998]. CJD is a rare disease with an estimated incidence rate of about one new case per million people each year. Higher rates have been reported among Libyan Jews (26 cases per million), and spatial clusters have been identified in areas of Slovakia, Hungary, England, U.S., and Chile [Johnson and Gibbs, Jr., 1998]. There are four different forms of CJD:

> *Sporadic CJD* (sCJD) refers to those cases in which there is no known infectious source and no evidence of the disease in the prior or subsequent generations of the patient's family. This is the most common subtype, corresponding to about 85% of cases. The cause of sCJD remains unknown, and there is no evidence of a causal link with animal TSEs. Patients are usually between 50 and 75 years old, and typical clinical features include a rapidly progressive dementia and myoclonus followed by the loss of the ability to move or speak [Brown et al., 1986; Collinge, 2001; Weber, 2000]. The first symptoms are subtle lapses of memory for daily events, changes in sleeping and eating behavior, and loss of interest and involvement in social activities. At this point, the illness may be confused as a mild depression. However, within a few weeks other features appear and develop quickly, including instability and hesitancy in walking, deteriorating vision, slowing of speech, mild or severe hallucinations, and dementia [Brown et al., 1986; Collinge, 2001; Weber, 2000]. After the first phase, the clinical abnormalities progress very rapidly, with development of urine incontinence, jerky movements, shakiness, stiffness of the limbs, and partial or complete loss of the ability to move. The disease is 100% fatal, and in most of the cases death occurs within a year after the appearance of the first clinical symptoms [Brown et al., 1986; Collinge, 2001; Johnson and Gibbs, Jr., 1998; Weber, 2000].

> *Familial CJD* (fCJD) is an inherited disease that represents approximately 10 to 15% of the total CJD cases. In most of these kindreds, point mutations, deletions, or insertions are found in the coding sequence of the prion protein gene [Kovacs et al., 2002; Prusiner and Scott, 1997]. More than 30 mutations in this gene have been described that are associated with phenotypes mimicking typical CJD or that induce distinctive progressive diseases with spongiform changes in the nervous system [Kovacs et al., 2002; Prusiner and Scott, 1997]. In general, fCJD has an earlier age of onset and a more prolonged course than the sporadic disease. The neuropathological abnormalities may vary in topographic distribution and in the prevalence of amyloid plaques, but the essential changes of vacuolization of neural cells with gliosis and neuronal loss are generally present.

> *Iatrogenic CJD* (iCJD) represents less than 5% of the cases and results from transmission of the causative agent via medical or surgical interventions using accidentally contaminated materials [Brown et al., 2000; Will, 2003]. Iatrogenic transmission of CJD has occurred in cases involving corneal transplants, implantation of electrodes in the brain, dura mater grafts, contaminated surgical instruments, and treatment with human growth hormone derived from cadaveric pituitaries [Brown et al., 2000; Will, 2003]. Although CJD can be

horizontally transmitted, the low incidence of iatrogenic CJD indicates that the disease is not contagious in the traditional sense. Family members or medical professionals who live with or treat CJD patients do not have a greater risk of getting the disease than the general population. The first case of iatrogenic CJD was reported in 1974 in the recipient of a corneal graft from a donor who had died of CJD. The highest number of iCJD cases has been seen in recipients of cadaveric pituitary human growth hormone (HGH) and dura mater grafts [Brown et al., 2000; Will, 2003]. The first case of iatrogenic CJD transmitted through HGH was reported in 1985, and since then the number of cases has increased steadily, reaching 139 cases in the year 2000 [Brown et al., 2000].

Variant CJD (vCJD) has appeared only recently and, in reality, corresponds to a different disease. Due to the enormous implications for human public health, the features of this disease will be described separately in this chapter.

In addition to CJD, three other clinically or pathologically similar prion diseases have been recognized in humans, including kuru, Gerstmann-Straussler-Scheinker (GSS) disease, and fatal insomnia. Several hundred cases of kuru have been reported, all confined to the Fore district in Papua, New Guinea [Liberski and Gajdusek, 1997]. A progressive cerebellar ataxia, uncoordinated movements, neurological weakness, palsies, and decay in brain cortical function characterize the disease [Liberski and Gajdusek, 1997]. Most kuru patients do not develop dementia, and this is a major clinical difference between kuru and CJD. In the late 1950s the average incidence of kuru in the Fore district was approximately 1%, but it could be as high as 10% in some villages. Approximately 67% of the kuru patients were adult women, 23% were children and adolescents, and only 10% were adult men [Liberski and Gajdusek, 1997]. Kuru was the first human TSE that was experimentally transmitted following intracerebral inoculation of brain homogenate into chimpanzees [Gajdusek et al., 1966]. Within the highlander population, the disease was transmitted during ritual cannibalism and has been virtually eliminated since the cessation of the handling and eating of the brains and other tissues of deceased relatives.

GSS is an autosomal dominant illness characterized by severe cerebellar ataxia and spastic paraparesis, with dementia developing late in the course of the disease [Ghetti et al., 1995]. The clinical course is much more prolonged than in CJD or other forms of TSE. The mean duration is approximately 5 years, with onset usually in either the third or fourth decade. Clinically speaking, GSS is characterized by dementia, ataxia, and sometimes seizures. Diagnosis is established by clinical examination and genetic screening for prion protein (PrP) mutations [Ghetti et al., 1995]. Histologically, the hallmark feature is the presence of multicentric PrP-amyloid plaques throughout the brain [Ghetti et al., 1995]. The most frequent mutation associated with

GSS is at codon 102 of the prion gene, but the syndrome has also been associated with mutations at other sites.

Fatal insomnia is inherited in most of the cases, but sporadic forms have recently been described [Scaravilli et al., 2000]. The illness is characterized by progressive insomnia, dysautonomia, and dementia, leading to death in 7 to 15 months [Cortelli et al., 1999]. Neuropathological features include selective atrophy of the ventral and mediodorsal thalamic nuclei, and in some patients spongiform changes are found.

1.2 Animal diseases

TSEs are known to affect various animal species, including sheep, goats, mink, mule deer, cows, cats, and exotic felines and ungulates [Collinge, 2001]. The most common animal TSE is scrapie, a disorder of sheep and goats that was first recognized in 1730 and became an endemic problem in several countries [Detwiler, 1992]. Its name comes from the observation that affected animals rub against the fences, scraping off their wool or hair to alleviate an intense itching. Later, the animals become ataxic, irritable, uninterested, and stop feeding. Examination of the brains of affected sheep after death shows many of the same changes that are found in people with CJD. There are some strains of sheep that seems particularly resistant to the development of scrapie (e.g., Scottish blackface), whereas others are more prone to it (e.g., herdwick, Suffolk) [Hunter et al., 1997]. Transmissibility was accidentally but stunningly demonstrated in 1937 when a population of Scottish sheep was inoculated against a common virus with a formalin extract of brain tissue unknowingly derived from an animal with scrapie [Cullie and Chelle, 1939]. After two years, nearly 10% of the flock developed scrapie. This finding constituted the first demonstration of the transmissibility of a prion disease and led to an enormous amount of work on the propagation of TSE. Although scrapie has been successfully transferred to several species of animals and might be the cause of some other animal TSE, there is no evidence of transmission of scrapie from sheep to humans [Hoinville, 1996].

Bovine spongiform encephalopathy (BSE) is a disease of cows that was first described in the U.K. by Wells and Wilesmith [Wilesmith et al., 1988] in 1986, and over 180,000 cases have been reported since then in that country alone (Figure 1.1A) [Kao et al., 2002]. The animals affected become uncoordinated and aggressive and lose weight. The number of cases in several countries of continental Europe is still increasing, and it is now clear that BSE is a pan-European problem affecting principally countries like Ireland, Portugal, France, Switzerland, Germany, Spain, Italy, Belgium, and the Netherlands (Figure 1.1B) [Bradley, 2002; Kao et al., 2002]. Small numbers of cases have been reported in the U.S., Canada, Japan, Slovakia, Czech Republic, Greece, Luxembourg, and Oman. The source of the BSE epidemic is probably food supplements that were prepared with bone meal from dead sheep [Wilesmith et al., 1988]. In 1988, the U.K. banned the use of these supplements

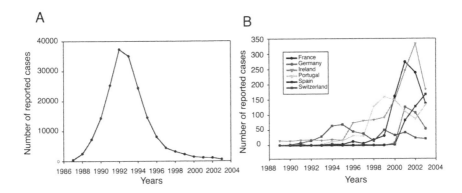

Figure 1.1 (See color insert after p. 114.) Cases of bovine spongiform encephalopathy reported in the U.K. (A) and some other European countries (B). (Information was obtained from the Web page of the European Intergovernmental Organization [http://oie.int/info/en_esbru. htm].)

in the preparation of cattle feed, and the number of cases has been diminishing after reaching a peak in 1992 (Figure 1.1A). However, in other countries the number of BSE cases is steady or even increasing (Figure 1.1B).

Chronic wasting disease (CWD) is a clinical syndrome affecting mule deer for more than 30 years [Williams and Miller, 2002]. Its origin is unknown. CWD-affected deer and elk show loss of body condition, changes in behavior, ataxia, head tremors, and somnolence. CWD can reduce the growth and size of wild deer and elk populations in areas where the prevalence is high, and it is of increasing concern for wildlife managers across North America [Williams and Miller, 2002] The disease was long thought to be limited in the wild to a relatively small endemic area in northeastern Colorado, southeastern Wyoming, and southwestern Nebraska, but it has recently been found in new areas of Colorado and Nebraska, as well as in wild deer in New Mexico, South Dakota, Wisconsin, and Saskatchewan [Williams and Miller, 2002]. At present, the risk of transmission of CWD to other animal species or to humans is unknown.

1.3 Variant CJD

Variant CJD (vCJD) is a new disease that was first described in March 1996 [Will et al., 1996]. In contrast to typical cases of sCJD, this variant form affects young patients (average age 27 years old) and has a relatively long duration of illness (median 14 months vs. 4.5 months in traditional CJD). Pathologically, vCJD demonstrates a consistent but previously unseen pattern [Collinge, 1999; Ironside et al., 1996]. More than 150 cases have been identified so far in the U.K., and a very small number of cases have been reported in other countries, including France, Ireland, Italy, Canada, Hong Kong, Spain, and the U.S. vCJD patients usually experience psychiatric symptoms early in the illness, which most commonly take the form of depression or, less often, a schizophrenia-like psychosis [Will et al., 2000]. Painful and persistent sensory abnormalities have been experienced by half of the cases early in the illness. Neurological signs, including ataxia (unsteadiness) and involuntary movements, develop as the illness progresses and, shortly before death, patients become completely immobile and mute [Ironside et al., 1996]. So far, all vCJD patients have a particular polymorphism for the gene codifying for the prion protein, in which the amino acid at position 129 is a methionine in both gene alleles [Andrews et al., 2003].

The hypothesis of a link between vCJD and BSE was first raised because of the association of these two TSEs in place and in time. Evidence supporting a link includes identification of pathological features similar to vCJD in brains of macaques inoculated with BSE [Lasmezas et al., 1996]. In addition, it has been demonstrated that vCJD is associated with a prion protein profile, as analyzed by Western blot studies, that distinguishes it from other forms of CJD and which resembles that seen in BSE transmitted to a number of other species [Collinge et al., 1996]. The most powerful evidence comes from studies showing that the transmission characteristics of BSE and vCJD in mice are almost identical [Bruce et al., 1997; Scott et al., 1999], strongly indicating that they are due to the same causative agent. It is also noteworthy that transgenic mice carrying a human gene have now been shown to be susceptible to BSE. Furthermore, no other plausible hypothesis for the occurrence of vCJD has been proposed, and intensive CJD surveillance in other European countries, with a low potential exposure to the BSE agent, has failed to identify additional cases. In conclusion, the most likely cause of vCJD is exposure to the BSE agent, plausibly due to dietary contamination by affected bovine central nervous system tissue.

Insufficient information is available at present to make any well-founded prediction about the future number of vCJD cases [Balter, 2001; Ghani, 2003]. Estimations have been done based on data about the experimental transmission of BSE to animals and the incubation times obtained in the transmission of kuru and iCJD. The transmission of these diseases does not involve a species barrier, and the mean incubation periods are approximately 10 to 15 years, suggesting that mean incubation periods of BSE in humans might be 30 years or more. Computational modeling studies using these parameters estimate the number of cases ranging from few hundred to 100,000 cases [Boelle et al., 2003; Cousens et al., 1997; d'Aignaux et al., 2001; Valleron et al., 2001]. Based on the recent decline in the number of new cases, it has been proposed that the epidemy is residing with a peak reached in the year 2000 [Andrews et al., 2003]. However, this analysis predicts an incubation period of 8 years, as calculated from the time in which the peaks for the BSE and vCJD epidemics were observed. Based on the knowledge of incubation times in the transmission of other forms of human prion diseases, it is highly unlikely that the incubation time in the cattle-to-human transmission can be so short. An alternative explanation is that the cases of vCJD that have so far occurred are part of the initial low number of people infected before the public onset of BSE and that the real vCJD epidemy has not yet appeared.

Considerable concern has been expressed about a possible iatrogenic transmission of vCJD from asymptomatic people incubating the disease through organ donation, surgery, or blood transfusion. One of the most debated issues is the possibility of transmission by blood or blood products. Several studies have shown that the disease can be propagated by blood in experimental rodents (For a review, see [Brown et al., 2001].) In addition, scrapie was shown transmissible through blood transfusion from sheep to sheep [Houston et al., 2000]. However, reports of infectivity in blood from patients with classical CJD are infrequent and have been questioned [Brown et al., 2001]. Interestingly, a recent study reported the first possible case of vCJD acquired by blood transfusion [Llewelyn et al., 2004]. In that report, a 69-year-old person developed vCJD symptoms 6.5 years after receiving a transfusion of red cells donated by an individual who made the donation 3.5 years before developing symptoms of vCJD. Furthermore, a case of potential preclinical vCJD was recently reported in a patient who died from a nonneurological disorder 5 years after receiving a blood transfusion from a donor who subsequently developed vCJD [Peden et al., 2004]. Protease-resistant prion protein (a marker for TSE diagnosis, see Chapter 8) was detected by Western blot and immunohistochemistry in the spleen, but not in the brain. The patient was heterozygote at codon 129 of the prion gene, suggesting that susceptibility to vCJD infection may not be restricted to the methionine homozygous individuals. If the possibility of transmission of vCJD by blood transfusion is supported by more cases like these, it could lead to dramatic consequences for the vCJD epidemic, because it would indicate that blood carries infectivity several years before the onset of clinical symptoms. Another route of transmission of vCJD could be via contaminated surgical instruments, as normal hospital sterilization procedures are probably not completely inactivating the prions [Frosh et al., 2001]. The extensive lymphoreticular involvement in vCJD [Wadsworth et al., 2001], which is likely to be present from a relatively early preclinical stage, raises the possibility that instruments could be contaminated in particular during procedures that involve contact with lymphoreticular tissues. This includes the common procedures of tonsillectomy, appendectomy, and lymph node and gastrointestinal biopsy. Recent studies have demonstrated that prions can adhere easily to metal surfaces, and prion-contaminated metal wires are efficient vehicles for experimental transmission of prion disease [Flechsig et al., 2001].

1.4 Neuropathology

In most of the TSE cases, there are no recognizable gross abnormalities in the brain. However, patients who survive for several years show variable degrees of cerebral atrophy comparable with those observed in other neurodegenerative diseases. The typical microscopic features of TSEs are vacuolation of the neuropil in the gray matter, prominent neuronal loss, exuberant reactive astrogliosis, and a variable degree of cerebral accumulation of prion protein aggregates (Figure 1.2) [Budka et al., 1995; MacDonald et al., 1996; Wells, 1993]. The most specific of these abnormalities is the vacuolation, giving to the brain the appearance of a sponge and hence the name of "spongiform encephalopathies." The spongiform degeneration consists of diffuse or focally clustered, small, round vacuoles that may become confluent [Wells, 1993]. The contribution of the vacuolization process to the pathogenesis of the disease is unclear. Neuronal loss occurs by the process of programmed cell death (usually termed "apoptosis") and is the most likely cause of brain malfunction [Gray et al., 1999]. Brain inflammation is prominent in TSEs, but it is rather unusual, because it consists mainly of activation of brain cells (astrocytes and microglia) [Betmouni et al., 1996]. All of the known prion diseases in animals and humans present accumulations of abnormal prion protein aggregates in the central nervous system, sometimes in the form of amyloid plaques, similar to the lesions found in other neurodegenerative illnesses, such as Alzheimer's disease [Ghetti et al., 1996; Jeffrey et al., 1995].

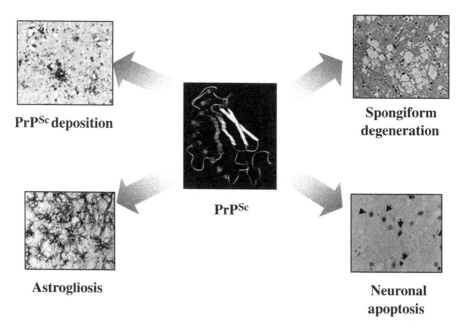

PrP^{Sc} deposition

Spongiform degeneration

PrP^{Sc}

Astrogliosis

Neuronal apoptosis

Figure 1.2 (See color insert after p. 114.) Neuropathological features of TSE. The central process in prion diseases appears to be the formation of the misfolded prion protein (PrP^{Sc}) that, in some cases, aggregates to form cerebral amyloid deposits. Through a mechanism that is not entirely known, this process induces neuronal apoptosis, spongiform brain degeneration, and astrogliosis.

A characteristic feature of TSE neuropathology is the large degree of variation in the distribution and magnitude of the abnormalities [Budka et al., 1995; Jeffrey et al., 1995]. These variations have been attributed to several factors, including the kinetics of prion propagation, cell tropism, differential toxicity, host neuroanatomy, and genetic makeup. For instance, sCJD rarely presents extensive accumulation of protein aggregates. In fCJD, amyloid plaques are sometimes found in the cerebellum adjacent to Purkinje cells and occasionally are grouped in arrays. In vCJD, numerous plaques are observed surrounded by a halo of intense vacuolation, giving them a distinctive appearance usually referred to as florid plaques [Ironside, 1998]. In iCJD, plaques are distributed more diffusely throughout the brain. In kuru, around 70% of cases have plaques, and this number increases in long-latency cases. In GSS, amyloid plaques are always present and have a unique multicentric configuration [Ghetti et al., 1995]. In fatal insomnia, no plaques have been observed. A similar spectrum of pathological differences is observed in the animal prion diseases [Wells, 1993].

1.5 Concluding remarks

TSEs are a group of diseases affecting humans and mammals. Despite their different etiology, clinical presentation, and profile of neuropathological alterations, they are grouped together because the central event in the pathogenesis is the same: the misfolding of the prion protein. In humans, TSEs are rare diseases; however, they have gained significant public attention due to their unprecedented mechanism of propagation and the possibility of transmission from animal food. Millions of people have been exposed to contaminated meat, and the extent of a possible vCJD epidemic remains unclear. The inexorable fatal outcome and dramatic clinical manifestation of the disease, added to the lack of any form of treatment and early diagnosis, contributes to the perception of TSEs as horrible diseases.

References* **

Andrews, N.J. et al., Deaths from variant Creutzfeldt-Jakob disease in the U.K., *Lancet*, 361, 751–752, 2003.

Balter, M., Infectious diseases: uncertainties plague projections of vCJD toll, *Science*, 294, 770–771, 2001.

Betmouni, S., Perry, V.H., and Gordon, J.L., Evidence for an early inflammatory response in the central nervous system of mice with scrapie, *Neuroscience*, 74, 1–5, 1996.

Boelle, P.Y. et al., Modelling the epidemic of variant Creutzfeldt-Jakob disease in the U.K. based on age characteristics: updated, detailed analysis, *Stat. Methods Med. Res.*, 12, 221–233, 2003.

* Highlight primary articles of outstanding importance and quality, including a small description of the findings.
** Highlight comprehensive review articles similar to the topic of this chapter.

Bradley, R., Bovine spongiform encephalopathy: update, *Acta Neurobiol. Exp.* (*Wars.*), 62, 183–195, 2002.

Brown, P. et al., Creutzfeldt-Jakob disease: clinical analysis of a consecutive series of 230 neuropathologically verified cases, *Ann. Neurol.*, 20, 597–602, 1986.

Brown, P., Cervenakova, L., and Diringer, H., Blood infectivity and the prospects for a diagnostic screening test in Creutzfeldt-Jakob disease, *J. Lab. Clin. Med.*, 137, 5–13, 2001.

Brown, P. et al., Iatrogenic Creutzfeldt-Jakob disease at the millennium, *Neurology*, 55, 1075–1081, 2000.

'Bruce, M.E. et al., Transmissions to mice indicate that "new variant" CJD is caused by the BSE agent, *Nature*, 389, 498–501, 1997. (Reports compelling evidence that vCJD is derived from BSE.)

Budka, H. et al., Neuropathological diagnostic criteria for Creutzfeldt-Jakob disease (CJD) and other human spongiform encephalopathies (prion diseases), *Brain Pathol.*, 5, 459–466, 1995.

Collinge, J., Variant Creutzfeldt-Jakob disease, *Lancet*, 354, 317–323, 1999.

"Collinge, J., Prion diseases of humans and animals: their causes and molecular basis, *Annu. Rev. Neurosci.*, 24, 519–550, 2001.

Collinge, J. et al., Molecular analysis of prion strain variation and the aetiology of "new variant" CJD, *Nature*, 383, 685–690, 1996.

Cortelli, P. et al., Fatal familial insomnia: clinical features and molecular genetics, *J. Sleep Res.*, 8 (supp. 1), 23–29, 1999.

Cousens, S.N. et al., Predicting the CJD epidemic in humans, *Nature*, 385, 197–198, 1997.

Cullie, J. and Chelle, P.L., Experimental transmission of trembling to the goat, *Comptes Rendus des Seances de l'Academie des Sciences*, 208, 1058–1160, 1939.

d'Aignaux, J.N., Cousens, S.N., and Smith, P.G., Predictability of the U.K. variant Creutzfeldt-Jakob disease epidemic, *Science*, 294, 1729–1731, 2001.

Detwiler, L.A., Scrapie, *Rev. Sci. Tech.*, 11, 491–537, 1992.

Flechsig, E. et al., Transmission of scrapie by steel-surface-bound prions, *Mol. Med.*, 7, 679–684, 2001.

Frosh, A., Joyce, R., and Johnson, A., Iatrogenic vCJD from surgical instruments, *BMJ*, 322, 1558–1559, 2001.

'Gajdusek, D.C., Gibbs, C.J., and Alpers, M., Experimental transmission of a kuru-like syndrome to chimpanzees, *Nature*, 209, 794–796, 1966. (First demonstration of transmission of a human prion disease.)

Ghani, A.C., Predicting the unpredictable: the future incidence of variant Creutzfeldt-Jakob disease. *Int. J. Epidemiol.*, 32, 792–793.

Ghetti, B. et al., Gerstmann-Straussler-Scheinker disease and the Indiana kindred, *Brain Pathol.*, 5, 61–75, 1995.

Ghetti, B. et al., Prion protein amyloidosis, *Brain Pathol.*, 6, 127–145, 1996.

Gray, F. et al., Neuronal apoptosis in Creutzfeldt-Jakob disease, *J. Neuropathol. Exp. Neurol.*, 58, 321-328, 1999.

Hoinville, L.J., A review of the epidemiology of scrapie in sheep, *Rev. Sci. Tech.*, 15, 827–852, 1996.

'Houston, F. et al., Transmission of BSE by blood transfusion in sheep, *Lancet*, 356, 999–1000, 2000. (Reports the transmission of the BSE agent from sheep to sheep by blood transfusion.)

Hunter, N. et al., Natural scrapie and PrP genotype: case-control studies in British sheep, *Vet. Rec.*, 141, 137–140, 1997.

Ironside, J.W., Neuropathological findings in new variant CJD and experimental transmission of BSE, *FEMS Immunol. Med. Microbiol.*, 21, 91–95, 1998.

Ironside, J.W. et al., A new variant of Creutzfeldt-Jakob disease: neuropathological and clinical features, *Cold Spring Harb. Symp. Quant. Biol.*, 61, 523–530, 1996.

Jeffrey, M., Goodbrand, I.A., and Goodsir, C.M., Pathology of the transmissible spongiform encephalopathies with special emphasis on ultrastructure, *Micron.*, 26, 277–298, 1995.

**Johnson, R.T. and Gibbs, C.J., Jr., Creutzfeldt-Jakob disease and related transmissible spongiform encephalopathies, *N. Engl. J. Med.*, 339, 1994–2004, 1998.

Kao, R.R. et al., The potential size and duration of an epidemic of bovine spongiform encephalopathy in British sheep, *Science*, 295, 332–335, 2002.

Kovacs, G.G. et al., Mutations of the prion protein gene phenotypic spectrum, *J. Neurol.*, 249, 1567–1582, 2002.

Lasmezas, C.I. et al., BSE transmission to macaques, *Nature*, 381, 743–744, 1996.

Liberski, P.P. and Gajdusek, D.C., Kuru: forty years later, a historical note, *Brain Pathol.*, 7, 555–560, 1997.

*Llewelyn, C.A. et al., Possible transmission of variant Creutzfeldt-Jakob disease by blood transfusion, *Lancet*, 363, 417–421, 2004. (Reports a case of vCJD with potential origin on a blood transfusion from an individual incubating the disease.)

MacDonald, S.T., Sutherland, K., and Ironside, J.W., Prion protein genotype and pathological phenotype studies in sporadic Creutzfeldt-Jakob disease, *Neuropathol. Appl. Neurobiol.*, 22, 285–292, 1996.

Peden, A.H. et al., Preclinical vCJD after blood transfusion in a *Prnp* codon 129 heterozygous patient, *Lancet*, 264, 527–529, 2004.

Prusiner, S.B., Prions, *Proc. Natl. Acad. Sci. USA*, 95, 13363–13383, 1998.

Prusiner, S.B. and Scott, M.R., Genetics of prions, *Annu. Rev. Genet.*, 31, 139–175, 1997.

Scaravilli, F. et al., Sporadic fatal insomnia: a case study, *Ann. Neurol.*, 48, 665–668, 2000.

Scott, M.R. et al., Compelling transgenetic evidence for transmission of bovine spongiform encephalopathy prions to humans, *Proc. Natl. Acad. Sci. USA*, 96, 15137–15142, 1999.

Valleron, A.J. et al., Estimation of epidemic size and incubation time based on age characteristics of vCJD in the United Kingdom, *Science*, 294, 1726–1728, 2001.

Wadsworth, J.D. et al., Tissue distribution of protease resistant prion protein in variant Creutzfeldt-Jakob disease using a highly sensitive immunoblotting assay, *Lancet*, 358, 171–180, 2001.

Weber, T., Clinical and laboratory diagnosis of Creutzfeldt-Jakob disease, *Clin. Neuropathol.*, 19, 249–250, 2000.

Wells, G.A., Pathology of nonhuman spongiform encephalopathies: variations and their implications for pathogenesis, *Dev. Biol. Stand.*, 80, 61–69, 1993.

Wilesmith, J.W. et al., Bovine spongiform encephalopathy: epidemiological studies, *Vet. Rec.*, 123, 638–644, 1988.

Will, R.G., Acquired prion disease: iatrogenic CJD, variant CJD, kuru, *Br. Med. Bull.*, 66, 255–265, 2003.

*Will, R.G. et al., A new variant of Creutzfeldt-Jakob disease in the U.K., *Lancet*, 347, 921–925, 1996. (First report of variant CJD.)

Will, R.G. et al., Diagnosis of new variant Creutzfeldt-Jakob disease, *Ann. Neurol.*, 47, 575–582, 2000.
Williams, E.S. and Miller, M.W., Chronic wasting disease in deer and elk in North America, *Rev. Sci. Tech.*, 21, 305–316, 2002.

chapter two

The infectious agent and the prion hypothesis

The infectious origin for transmissible spongiform encephalopathy (TSE) was accidentally discovered in 1937 when a population of Scottish sheep was inoculated against a common virus with a formalin extract of brain tissue unknowingly derived from an animal with scrapie. Scrapie was subsequently transmitted experimentally to sheep [Cullie and Chelle, 1939] and later to mice [Chandler, 1961]. In humans, an infectious origin was suspected as a route of propagation of kuru among the cannibalistic tribes of New Guinea, and this was demonstrated in 1966 by transmission of kuru to monkeys in the pioneering studies of Carleton Gajdusek [Gajdusek et al., 1966]. These studies were followed by transmission of Creutzfeldt-Jakob disease (CJD) to animals [Gibbs, Jr., et al., 1968] and, interestingly, a familial form of TSE, Gerstmann-Straussler-Scheinker (GSS) syndrome [Masters et al., 1981]. This chapter describes the various hypotheses proposed for the infectious agent and, in particular, the compelling evidence for the prion hypothesis as well as the criticisms and missing evidence for this novel type of infectious agent.

2.1 Hypothesis for the infectious agent

The nature of the transmissible agent is perhaps the most extensively studied aspect of TSE research [Soto and Castilla, 2004]. (See Table 2.1 for a historical outline of major events in this subject.) Despite this, it remains a matter of passionate controversy [Chesebro, 1998; Mestel, 1996]. Initially, the agent was thought to be a slow virus (Figure 2.1) because of the unusually long incubation period between the time of exposure to the pathogen and the onset of symptoms [Cho, 1976]. However, further research has indicated that this agent differs significantly from viruses and other conventional agents. In 1966, Alper and colleagues reported that the minimum molecular weight of the scrapie agent that still maintained infectivity was too small ($\approx 2 \times 10^5$) to possibly be a virus or any other known type of infectious agent [Alper

Table 2.1 Outline of the Most Relevant Milestone Related to TSE Infectivity and the Nature of the Infectious Agent

Year	Description
1937	Evidence of the transmissible nature of TSE in sheep inoculated with a vaccine prepared from infected lymphoid tissue
1939	Experimental transmission of scrapie
1961	Scrapie experimentally transmitted to mice
1966	Kuru transmitted to chimpanzees
1966	Demonstration of small size of scrapie agent
1967	Scrapie agent shown to be highly resistant to DNA destruction
1967	First enunciation of the protein-only hypothesis
1980	Protease-resistant and hydrophobic protein discovered in scrapie-infected hamster brain
1981	A hereditary human disease (GSS) is transmitted to animals
1982	Prion concept enunciated
1982	Purification and characterization of prion protein
1984	Propagation of infectivity in cells
1985	Gene encoding PrP cloned
1988	Infectivity neutralized by anti-PrP antibodies
1989	First mutation in PrP gene associated to familial TSE
1990	Production of transgenic animal expressing mutant PrP that develop clinical and pathological signs of TSE
1993	PrP knock out mice is resistant to prion infection
1993	Conformational differences between PrP^C and PrPres reported
1994	Cell-free conversion of PrP^C into PrPres
1994	Expansion of the prion concept to yeast proteins
1996	Mutant PrP produced in cells exhibit some biochemical properties of PrPres but lacks infectivity
2000	*In vivo* transmission of yeast Sup35 prion generated *in vitro*
2001	High-efficiency *in vitro* conversion of PrP^C into PrPres by PMCA
2004	*In vivo* propagation of different yeast prion strains generated from recombinant protein *in vitro*
2004	First report of *in vitro* produced synthetic mammalian prions
2005	Demonstration of *in vitro* propagation of infectious prions

et al., 1966]. A year later, the same group demonstrated that the infectious agent was extremely resistant to treatments that normally destroy nucleic acids, such as ultraviolet (UV) and ionizing radiation [Alper et al., 1967]. In addition, over the past 30 years, many research groups throughout the world have unsuccessfully attempted to find a virus associated with the disease [Prusiner, 1998]. These and other results led some investigators to propose a modification of the viral theory that became known as the virino hypothesis (Figure 2.1). A virino can be defined as a small informational molecule (most likely a nucleic acid) associated with a host protein that serves as a coat [Carp et al., 1994; Kimberlin, 1982]. However, this proposal has not gained broad support, mainly because of the overwhelming amount of data in favor of the alternative "protein-only" hypothesis.

The "protein-only" hypothesis, first enunciated by J.S. Griffith in 1967 to explain Alper's findings, proposed that the material responsible for the

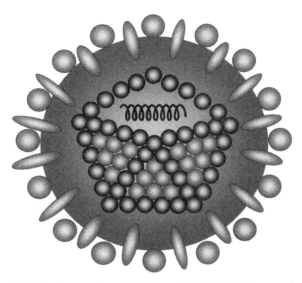

VIRUS. A conventional infectious agent consisting
of a protein-lipid capsule and a nucleic acid genome

VIRINO. Infectious agent composed of a highly-
resistant and compact protein coat covering a small
piece of an informational molecule

PRION. An infectious protein that replicates by
transferring protein misfolding in the absence of
nucleic acid

Figure 2.1 Hypothesis for the nature of the infectious agent associated with TSE. Although many hypotheses have been proposed over the years, the most seriously considered are the viral, virino, and protein-only hypothesis.

disease transmission was uniquely a protein that has the surprising ability to replicate itself in the body [Griffith, 1967]. Since the 1980s, the protein-only hypothesis of TSE transmissibility has been led by Stanley Prusiner's group, who coined the name "prion" for this novel proteinaceous infectious particle and has produced some of the most important evidence to support it.

2.2   Evidence supporting the prion hypothesis

A critical finding in understanding the nature of this novel infectious agent was the isolation of the protease-resistant and misfolded prion protein (PrP^{Sc}) from the infectious material [Bolton et al., 1982]. It was shown that PrP^{Sc} and infectivity copurified and that the concentration of the protein was proportional to the infectivity titer [Gabizon et al., 1988]. Moreover, infectivity was shown to be retained in highly purified preparations of PrP^{Sc} in which no other component was detectable. In addition, infectivity was convincingly reduced by agents that destroy protein structure and, more importantly, by anti-PrP antibodies [Gabizon et al., 1988]. Purification of PrP made it possible to identify the gene encoding PrP [Chesebro et al., 1985; Oesch et al., 1985]. PrP mRNA is the product of a single host gene that is present in the brain of uninfected animals and is constitutively expressed by many cell types. Thus, it became clear that PrP can exist in two alternative forms: the normal cellular protein (termed PrP^{C}) and the pathological isoform (termed PrP^{Sc}) (See Table 2.2 for an explanation of the nomenclature of PrP isoforms). Chemical differences to distinguish these two PrP isoforms have not been detected [Stahl et al., 1993], and the conversion seems to involve a conformational change whereby the α-helical content of the normal protein diminishes and the amount of β-sheet increases [Pan et al., 1993]. The structural changes are followed by alterations in other biochemical properties, such as protease resistance, solubility, and the ability to form larger-order aggregates [Cohen and Prusiner, 1998].

Table 2.2 Nomenclature for the Different Prion Protein Isoforms

PrP: Refers to the total prion protein without making a distinction for different isoforms.

PrP^{C}: Normal cellular prion protein present in healthy people. This form is rich in α-helical conformation, is soluble and protease-sensitive.

PrP^{Sc}: Disease-associated misfolded prion protein present in individuals affected by TSE. This form is infectious, rich in β-sheet conformation, insoluble and mostly protease-resistant.

PrP^{res}: Refers to a β-sheet rich, protease-resistant prion protein, which may or may not be identical to PrP^{Sc}. In particular, this name is used to refer to *in vitro* produced protease resistant protein, which has not been experimentally shown to be infections.

PrP27-30: Correspond to the protein core that remains resistant after protease treatment of PrP^{Sc} or PrP^{res}. It consists of the last two thirds of the protein

Another piece of evidence comes from genetic studies showing that most, if not all, of the familial cases of TSE are linked to mutations in the PrP gene [Collinge, 2001; Prusiner, 1998]. These findings not only provide support for a central role of PrP in the disease pathogenesis, but also are strong evidence for the protein-only hypothesis, because once the genetic disease arises, it can then be propagated in an infectious way. Interestingly, a TSE-like disease was produced in mice overexpressing PrP genes with point mutations linked to GSS disease [Hsiao et al., 1990]. These animals spontaneously develop neurological dysfunction, spongiform brain degeneration, and astrocytic gliosis, and the disease was transmitted to animals expressing the mutant genes [Hsiao et al., 1994]. However these results have been highly controversial and, indeed, another group has failed to observe disease in transgenic animals in which gene targeting was used to replace the mouse PrP gene by the mutant gene [Barron and Manson, 2003]. These data might be interpreted to suggest that a large overexpression of the PrP gene is toxic, leading to a disease phenotype regardless of the mutation or the transmissible origin.

Particularly strong evidence in favor of the prion hypothesis came from the group of Charles Weissmann [Bueler et al., 1993], who showed that mice devoid of the PrP gene were resistant to scrapie infection, developing neither signs of scrapie nor allowing propagation of the infectious agent. Although other groups showed that some PrP-null animals developed some neurological alterations, this was not due to the lack of PrP, but rather to the artifactual expression of another gene in the brain [Moore et al., 1999; Rossi et al., 2001].

Another important finding was the successful propagation of infectivity in neuroblastoma cells [Race et al., 1987; Rubenstein et al., 1984]. These cells can be chronically infected with brain homogenate containing PrP^{Sc}, and the infectious agent as well as the misfolded protein can be maintained over several months. A further milestone supporting the prion hypothesis was the cell-free conversion of PrP^{C} into a protease-resistant PrP^{Sc}-like isoform (termed PrP^{res}, see Table 2.2) catalyzed by the pathological protein. The original system, developed by Caughey and coworkers [Kocisko et al., 1994], using purified PrP^{C} mixed with stoichiometric amounts of purified PrP^{res}, has been very useful in understanding the mechanism of prion replication and in identifying compounds that interfere with this process (for review, see [Caughey, 2003]). However, the low yield of PrP^{res} formation and the nonphysiological conditions used have made difficult to evaluate the biological and structural properties of the newly converted protein. Nevertheless, the fact that PrP^{res} was able to induce the transformation of the normal protein into more PrP^{res} was important evidence in favor of the prion hypothesis. More recently, we have reported a new *in vitro* conversion system for transforming large quantities of PrP^{C} induced by minute amounts of PrP^{Sc} [Saborio et al., 2001]. This system, called PMCA (protein misfolding cyclic amplification), confirms a critical facet of the prion hypothesis, which is that prion replication is a cyclical process and that newly produced PrP^{res} is able to maintain the propagation of the misfolded protein [Saborio et al., 2001].

2.3   Criticisms of the prion hypothesis

Despite compelling evidence in favor of the prion hypothesis and the fact that it neatly explains many of the observed features of TSE, the hypothesis does have one particular weakness that has long been exploited as evidence against it [Chesebro, 1998]. Scrapie and other TSEs are known to exhibit various "strains" characterized by different incubation periods, clinical features, and neuropathology [Prusiner, 1998]. (See Chapter 5 for an in-depth description of the strain phenomenon.) In infectious diseases, different strains generally arise from mutations or polymorphisms in the genetic makeup of the infectious agent. To reconcile the infectious agent composed exclusively of a protein with the strain phenomenon, it has been proposed that PrP^{Sc} obtained from different prion strains has slightly different conformation or aggregation states that can faithfully replicate at the expense of the host PrP^{C} (Figure 2.2) [Prusiner, 1998]. These different states of PrP^{Sc} may have distinct capabilities of catalyzing PrP^{C} conversion and may selectively target different brain regions, leading to the diversity of clinical symptoms and neuropathological alterations characteristic of prion strains (Figure 2.2). Support for this concept has come from various studies showing that PrP^{res} from different strains has noticeably distinct secondary structures [Caughey et al., 1998; Safar et al., 1998] whose properties can faithfully be transferred to PrP^{C} *in vitro* [Bessen et al., 1995]. However, up to now it has not been shown whether such differences are the cause or simply another manifestation of the prion-strain phenomenon. In addition, in some species (such as mice) more than 20 strains have been reported, and it is difficult to imagine a protein able to adopt more than 20 different and stable foldings.

The presence and quantity of PrP^{res} usually correlates with infectivity [Gabizon et al., 1988; Prusiner, 1998]. However, there have been reports in which infectivity is propagated in the absence of detectable PrP^{res} [Lasmezas et al., 1997], and other studies have shown that samples containing abundant amounts of PrP^{res} have little or no infectivity [Hill et al., 2000]. To explain these results, prion proponents argue that, although protease resistance is a typical biochemical feature of the misfolded infectious protein, it does not necessarily equate with infectivity and, indeed, only part of the infectious protein is protease resistant [Safar et al., 1998].

Although it has not been possible to identify a nucleic acid consistently associated with infectious preparations of PrP^{Sc} [Prusiner, 1998], several groups have reported the presence of small quantities of nucleic acids in infectious samples. (For references, see [Chesebro, 1998; Narang, 2002].) Moreover, PrP^{res} interacts with high affinity with nucleic acids, especially RNA [Derrington et al., 2002; Weiss et al., 1997], and recently it has been shown that RNA may help to catalyze the conversion of PrP^{C} into PrP^{res} *in vitro* [Deleault et al., 2003].

Another argument often used against the prion hypothesis is the lack of infectious origin of other neurodegenerative and systemic disorders associated with protein misfolding and aggregation, such as Alzheimer's disease,

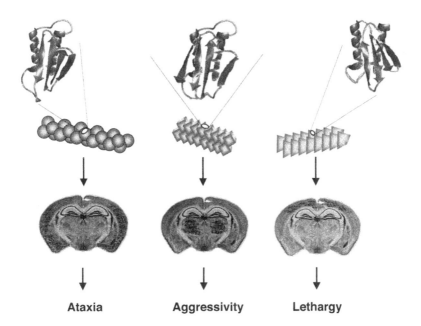

Ataxia   **Aggressivity**   **Lethargy**

Figure 2.2 The strain diversity might be encoded in different PrP^Sc conformations. The strain phenomenon has long been used as evidence against the protein-only hypothesis. However, recent data suggest that the strain diversity is due to different conformations of misfolded prion protein, which form different aggregates that faithfully replicate their properties and lead to different histological and clinical symptoms.

Parkinson's disease, and peripheral amyloidosis. (For a further description of these diseases, see Chapter 11.) In many of these diseases, interaction between normal and misfolded proteins leads to the formation of new pathological proteins by a process known as seeding-nucleation polymerization, strikingly similar to the PrP conversion mechanism (see Chapter 4) [Soto, 2003]. The apparent lack of infectivity of these other misfolded proteins suggests that the protein interactions and induced misfolding events common to these diseases might not be the explanation of the unique TSE transmissibility. However, it is yet not clear whether or not some of these diseases may, under certain conditions, be transmissible, and indeed, some intriguing evidence in this regard has been recently published (see Chapter 12) [Kane et al., 2000; Lundmark et al., 2002]. If transmissibility can be convincingly demonstrated with some of the other protein-misfolding diseases, this will actually turn into strong evidence in favor of the prion hypothesis.

2.4   In vitro *generation of prions*

It is widely accepted that the definitive proof for the protein-only hypothesis would be the generation of infectivity in the test tube [Soto and Castilla,

2004]. If the infectious agent is misfolded PrP^{Sc} and its replication is promoted by interaction with PrP^C, then it should be possible to reproduce the whole process entirely *in vitro*. Many groups have attempted this goal using a variety of different approaches [May et al., 2004; Soto and Castilla, 2004]. One strategy involved the production of PrP containing mutations associated with inherited TSEs. Several mutant PrP^{Sc}-like molecules have been generated, some of which have been shown to acquire various biochemical properties of PrP^{res}, but so far none of them has been shown to be infectious [Chiesa et al., 1998; Lehmann and Harris, 1996].

In another approach to proving the hypothesis, workers have attempted to induce misfolding of recombinant prion protein or short PrP synthetic peptides into β-sheet-rich structures that exhibit some of the biochemical and biological properties of PrP^{Sc} [Baskakov, 2004; De Gioia et al., 1994; Jackson et al., 1999; Lee and Eisenberg, 2003]. The hope in these experiments was that even if a very small percentage of the protein altered *in vitro* adopted the "infectious folding," infectivity should have been generated. Although many of these studies failed to create infectivity *de novo*, a recent article from Prusiner and coworkers has reported positive results [Legname et al., 2004]. In these studies, a recombinant mouse-PrP fragment (residues 89–230) assembled into amyloid fibrils was found to induce a transmissible spongiform encephalopathy with PrP^{res} formation when injected in transgenic mice overexpressing the same PrP sequence [Legname et al., 2004]. These findings come close to be the long-awaited definitive proof of the prion hypothesis, but some experimental caveats need to be addressed in future publications. The fact that the disease was originally transmitted to transgenic animals overexpressing the PrP gene and not to wild-type animals is a matter of concern, because it is well known that animals overexpressing the prion gene develop spontaneously a prionlike disease. In addition, the clinical and histopathological presentation of the disease was very different than that of the usual disease in mice. The authors argue that this result may be due to the creation of a new strain of prions. In any case, if infectivity was indeed generated, the efficiency of the process is extremely low, because out of the 15 μg of recombinant protein injected per animal, no more than 0.000006% of this was infectious. This estimation is based on the known concentrations of PrP^{res} in the brain and infectivity titration experiments by mouse bioassay.

Generation of infectivity by *in vitro* conversion of PrP^C has also been explored. In these experiments, the conversion process is triggered and catalyzed by brain-derived PrP^{Sc}, and hence the formation of the correct "infectious folding" is more likely to occur. In the cell-free system developed by Caughey and coworkers [Kocisko et al., 1994], the low yield of conversion made it difficult to distinguish the potentially newly generated infectivity from the vast amount of infectivity used to begin the conversion. However, Collinge and coworkers took advantage of the species-barrier phenomenon to test infectivity of newly generated PrP^{res} under conditions in which the PrP^{res} from the inoculum would not be infectious [Hill et al., 1999]. The results of these experiments argued that cell-free-generated PrP^{res} was not infectious.

PMCA provides a new opportunity for evaluating the infectious properties of PrP^{res} generated *in vitro* because, after amplification under optimal conditions, >99.99999% of protease-resistant protein is composed of newly produced PrP^{res}. The high yield following conversion is essential to distinguish newly generated infectivity from that used to initiate the reaction. Recently, Kretzschmar and coworkers reported that serial PMCA, yielding a 300-fold amplification of PrP^{res} after 100 PMCA cycles, showed a moderated increase in infectivity [Bieschke et al., 2004]. However, the infectivity raise was within the effect predicted for sonication alone, making it impossible to conclude accurately whether or not newly generated PrP^{res} was infectious. As described in Chapter 10, we have recently been able to generate infectivity by *in vitro* replication of PrP^{res} using an improved version of PMCA.

2.5   Concluding remarks

The infectious agent associated with TSEs has been the subject of intense investigation. The unprecedented characteristics of propagation and the unique chemical structure of the agent continue to puzzle scientists. The prevailing hypothesis proposes a radically novel form of infectious agent (termed a "prion"), composed exclusively of protein, that propagates in the absence of nucleic acid by transmitting its folding characteristics to equivalent protein molecules in the host that are folded in a biologically normal conformation. Prions defy several important and well-established dogmas in biology and have opened an entirely new way of understanding infectious agents and protein folding. For many years, the prion hypothesis was viewed with much skepticism, but much compelling evidence in support of this model has mostly settled the debate. The overall conclusion from these findings is that the prion hypothesis is most likely correct and that the only component necessary to carry prion infectivity is misfolded PrP^{Sc}.

References* **

Alper, T. et al., Does the agent of scrapie replicate without nucleic acid? *Nature*, 214, 764–766, 1967.

*Alper, T., Haig, D.A., and Clarke, M.C., The exceptionally small size of the scrapie agent, *Biochem. Biophys. Res. Commun.*, 22, 278–284, 1966. (This, as well as the previously cited article, represents the first indications of the unconventional nature of the TSE infectious agent.)

Barron, R.M. and Manson, J.C., A gene-targeted mouse model of P102L Gerstmann-Straussler-Scheinker syndrome, *Clin. Lab. Med.*, 23, 161–173, 2003.

---

* Highlights primary articles of outstanding importance and quality, including a short description of the findings.
** Highlights comprehensive review articles related to the topic of this chapter.

Baskakov, I.V., Autocatalytic conversion of recombinant prion proteins displays a species barrier, *J. Biol. Chem.*, 279, 7671–7677, 2004.

Bessen, R.A. et al., Non-genetic propagation of strain-specific properties of scrapie prion protein, *Nature*, 375, 698–700, 1995.

Bieschke, J. et al., Autocatalytic self-propagation of misfolded prion protein, *Proc. Natl. Acad. Sci. USA*, 101, 12207–12211, 2004.

*Bolton, D.C., McKinley, M.P., and Prusiner, S.B., Identification of a protein that purifies with the scrapie prion, *Science*, 218, 1309–1311, 1982. (Reports the identification of the prion protein as a major constituent of scrapie infectious agent, opening the door for the "protein only" hypothesis of prion propagation.)

*Bueler, H. et al., Mice devoid of PrP are resistant to scrapie, *Cell*, 73, 1339–1347, 1993. (This seminal work demonstrated that the presence of the normal prion protein is essential to enable propagation of the TSE infectious agent.)

Carp, R.I. et al., The nature of the scrapie agent: biological characteristics of scrapie in different scrapie strain-host combinations, *Ann. N.Y. Acad. Sci.*, 724, 221–234, 1994.

Caughey, B., Prion protein conversions: insight into mechanisms, TSE transmission barriers and strains, *Br. Med. Bull.*, 66, 109–120, 2003.

Caughey, B., Raymond, G.J., and Bessen, R.A., Strain-dependent differences in beta-sheet conformations of abnormal prion protein, *J. Biol. Chem.*, 273, 32230–32235, 1998.

Chandler, R.L., Encephalopathy in mice produced by inoculation with scrapie brain material, *Lancet*, 1, 1378–1379, 1961.

**Chesebro, B., BSE and prions: uncertainties about the agent, *Science*, 279, 42–43, 1998.

Chesebro, B. et al., Identification of scrapie prion protein-specific mRNA in scrapie-infected and uninfected brain, *Nature*, 315, 331–333, 1985.

Chiesa, R. et al., Neurological illness in transgenic mice expressing a prion protein with an insertional mutation, *Neuron*, 21, 1339–1351, 1998.

Cho, H.J., Is the scrapie agent a virus? *Nature*, 262, 411–412, 1976.

Cohen, F.E. and Prusiner, S.B., Pathologic conformations of prion proteins, *Annu. Rev. Biochem.*, 67, 793–819, 1998.

Collinge, J., Prion diseases of humans and animals: their causes and molecular basis, *Annu. Rev. Neurosci.*, 24, 519–550, 2001.

Cullie, J. and Chelle, P.L., Experimental transmission of trembling to the goat, *Comptes Rendus des Seances de l'Academie des Sciences*, 208, 1058–1160, 1939.

De Gioia, L. et al., Conformational polymorphism of the amyloidogenic and neurotoxic peptide homologous to residues 106-126 of the prion protein, *J. Biol. Chem.*, 269, 7859–7862, 1994.

Deleault, N.R., Lucassen, R.W., and Supattapone, S., RNA molecules stimulate prion protein conversion, *Nature*, 425, 717–720, 2003.

Derrington, E. et al., PrP^C has nucleic acid chaperoning properties similar to the nucleocapsid protein of HIV-1, *C.R. Acad. Sci. III*, 325, 17–23, 2002.

Gabizon, R. et al., Immunoaffinity purification and neutralization of scrapie prion infectivity, *Proc. Natl. Acad. Sci. USA*, 85, 6617–6621, 1988.

Gajdusek, D.C., Gibbs, C.J., and Alpers, M., Experimental transmission of a Kuru-like syndrome to chimpanzees, *Nature*, 209, 794–796, 1966.

Gibbs, C.J., Jr., et al., Creutzfeldt-Jakob disease (spongiform encephalopathy): transmission to the chimpanzee, *Science*, 161, 388–389, 1968.

*Griffith, J.S., Self-replication and scrapie, *Nature*, 215, 1043–1044, 1967. (This visionary article enunciates for the first time the protein-only hypothesis of TSE propagation.)

Hill, A.F., Antoniou, M., and Collinge, J., Protease-resistant prion protein produced *in vitro* lacks detectable infectivity, *J. Gen. Virol.*, 80 (pt. 1), 11–14, 1999.

Hill, A.F. et al., Species-barrier-independent prion replication in apparently resistant species, *Proc. Natl. Acad. Sci. USA*, 97, 10248–10253, 2000.

Hsiao, K.K. et al., Serial transmission in rodents of neurodegeneration from transgenic mice expressing mutant prion protein, *Proc. Natl. Acad. Sci. USA*, 91, 9126–9130, 1994.

Hsiao, K.K. et al., Spontaneous neurodegeneration in transgenic mice with mutant prion protein, *Science*, 250, 1587–1590, 1990.

Jackson, G.S. et al., Reversible conversion of monomeric human prion protein between native and fibrilogenic conformations, *Science*, 283, 1935–1937, 1999.

Kane, M.D. et al., Evidence for seeding of beta-amyloid by intracerebral infusion of Alzheimer brain extracts in beta-amyloid precursor protein-transgenic mice, *J. Neurosci.*, 20, 3606–3611, 2000.

Kimberlin, R.H., Scrapie agent: prions or virinos? *Nature*, 297, 107–108, 1982.

*Kocisko, D.A. et al., Cell-free formation of protease-resistant prion protein, *Nature*, 370, 471–474, 1994. (First report on *in vitro* conversion of PrP^C into PrP^{res}.)

Lasmezas, C.I. et al., Transmission of the BSE agent to mice in the absence of detectable abnormal prion protein, *Science*, 275, 402–405, 1997.

Lee, S. and Eisenberg, D., Seeded conversion of recombinant prion protein to a disulfide-bonded oligomer by a reduction-oxidation process, *Nat. Struct. Biol.*, 10, 725–730, 2003.

*Legname, G. et al., Synthetic mammalian prions, *Science*, 305, 673–676, 2004. (Describes for the first time the generation of infectious prion protein *in vitro*. Although several experimental concerns have to be addressed, this study represents an important step toward confirmation of the prion hypothesis.)

Lehmann, S. and Harris, D.A., Two mutant prion proteins expressed in cultured cells acquire biochemical properties reminiscent of the scrapie isoform, *Proc. Natl. Acad. Sci. USA*, 93, 5610–5614, 1996.

*Lundmark, K. et al., Transmissibility of systemic amyloidosis by a prionlike mechanism, *Proc. Natl. Acad. Sci. USA*, 99, 6979–6984, 2002. (An intriguing article showing a prion-like phenomenon in a systemic amyloidosis-related disorder.)

Masters, C.L., Gajdusek, D.C., and Gibbs, C.J., Jr., Creutzfeldt-Jakob disease virus isolations from the Gerstmann-Straussler syndrome with an analysis of the various forms of amyloid plaque deposition in the virus-induced spongiform encephalopathies, *Brain*, 104, 559–588, 1981.

May, B.C. et al., Prions: so many fibers, so little infectivity, *Trends Biochem. Sci.*, 29, 162–165, 2004.

Mestel, R., Putting prions to the test, *Science*, 273, 184–189, 1996.

Moore, R.C. et al., Ataxia in prion protein (PrP)-deficient mice is associated with upregulation of the novel PrP-like protein doppel, *J. Mol. Biol.*, 292, 797–817, 1999.

**Narang, H., A critical review of the nature of the spongiform encephalopathy agent: protein theory versus virus theory, *Exp. Biol. Med. (Maywood.)*, 227, 4–19, 2002.

Oesch, B. et al., A cellular gene encodes scrapie PrP 27-30 protein, *Cell*, 40, 735–746, 1985.

Pan, K.M. et al., Conversion of alpha-helices into β-sheets features in the formation of scrapie prion proteins, *Proc. Natl. Acad. Sci. USA*, 90, 10962–10966, 1993.

"Prusiner, S.B., Prions, *Proc. Natl. Acad. Sci. USA*, 95, 13363–13383, 1998.

Race, R.E., Fadness, L.H., and Chesebro, B., Characterization of scrapie infection in mouse neuroblastoma cells, *J. Gen. Virol.*, 68 (pt. 5), 1391–1399, 1987.

Rossi, D. et al., Onset of ataxia and Purkinje cell loss in PrP null mice inversely correlated with Dpl level in brain, *EMBO J.*, 20, 694–702, 2001.

'Rubenstein, R., Carp, R.I., and Callahan, S.M., *In vitro* replication of scrapie agent in a neuronal model: infection of PC12 cells, *J. Gen. Virol.*, 65 (pt. 12), 2191–2198, 1984. (First demonstration of the replication of TSE infectious agents in cell cultures.)

Saborio, G.P., Permanne, B., and Soto, C., Sensitive detection of pathological prion protein by cyclic amplification of protein misfolding, *Nature*, 411, 810–813, 2001.

Safar, J. et al., Eight prion strains have PrP(Sc) molecules with different conformations, *Nat. Med.*, 4, 1157–1165, 1998.

Soto, C., Unfolding the role of protein misfolding in neurodegenerative diseases, *Nature Rev. Neurosci.*, 4, 49–60, 2003.

"Soto, C. and Castilla, J., The controversial protein-only hypothesis of prion propagation, *Nature Med.*, 10, S63–S67, 2004.

Stahl, N. et al., Structural studies of the scrapie prion protein using mass spectrometry and amino acid sequencing, *Biochem.*, 32, 1991–2002, 1993.

Weiss, S. et al., RNA aptamers specifically interact with the prion protein PrP, *J. Virol.*, 71, 8790–8797, 1997.

chapter three

The prion protein: structure, conversion, and mechanism of propagation

The identification of (a) the pathological prion protein (PrP^{Sc}) in the brain of animals affected by scrapie, (b) the posterior localization of the gene, and (c) the normal version of the prion protein (PrP^C) led to a great amount of work to understand the differences between the two PrP isoforms and the mechanism by which the normal protein is converted into the disease-associated protein. This chapter describes the current knowledge about the structural properties of PrP isoforms and the mechanism by which prions propagate in the body by transmission of protein misfolding.

3.1 Structural features of the cellular and scrapie prion protein isoforms

PrP is the product of a single gene, called *Prnp*, that directs the synthesis of a 253-residue protein (252 residues in some species), which contains a signal peptide for secretion, five octapeptide repeats near the amino-terminus, two glycosylation sites, and one disulfide bridge [Cohen, 1999] (Figure 3.1). In addition, a glycosylphosphatidylinositol (GPI) anchor attaches the protein to the outer surface of the cell membrane. The mature protein is produced after removal of the N- and C-terminal signal sequences during its transit through the endoplasmic reticulum (ER) and the Golgi apparatus [Harris, 2003]. PrP is constitutively expressed in the brain and other tissues of healthy individuals as three glycoforms (mono-, di-, and unglycosylated), the proportions of which differ in different species and strains [Hill et al., 2003; Parchi et al., 1997]. No other posttranslational modifications have been systematically and reproducibly detected in PrP. Strikingly, the misfolded form contains the same posttranslational modifications as PrP^C, and extensive

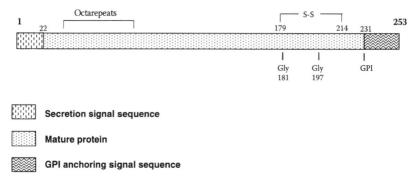

Figure 3.1 Schematic representation of the posttranslational modification in PrP. PrP^C is a 253-amino acid protein containing two signal peptides for export to the cell surface and for anchoring it at the membrane by addition of a GPI motif. It also has two glycosylation sites and an intramolecular cysteine bridge. At the N-terminus, it contains several octapeptide repeats that participate in the putative PrP biological activity of copper binding.

studies have failed to identify chemical differences between the two isoforms [Stahl et al., 1993]. Despite the similarity in primary structure, PrP^C and PrP^{Sc} show drastically different biochemical and physicochemical properties, such as protease resistance, solubility, resistance to denaturation, and aggregability [Cohen and Prusiner, 1998]. These differences are most likely due to the radical changes on the secondary and tertiary structure that feature the conversion of PrP^C into PrP^{Sc} [Caughey et al., 1991; Pan et al., 1993].

PrP^C consists of two structural domains; an unordered N-terminal fragment up to residue 128, which contains the octapeptide repeats and likely plays a role in PrP biological functions (see Chapter 4), and a globular C-terminal domain composed of three α-helices and two small β-strand regions (Figure 3.2). The tridimensional structure of PrP^C has been resolved by NMR (nuclear magnetic resonance) for several species of mammals, including human, mouse, hamster, cattle, sheep, etc. [Calzolai et al., 2005; Gossert et al., 2005; Lopez et al., 2000; Lysek et al., 2005; Riek et al., 1996; Zahn et al., 2000]. Strikingly, despite the presence of a number of amino acid substitutions, the tridimensional structure of PrP^C is highly conserved. Recently, the crystalline structure of a dimeric form of recombinant human PrP^C has been obtained by X-ray diffraction [Knaus et al., 2001]. The dimer is formed by the three-dimensional swapping of the third α-helix and rearrangement of the disulfide bond. A new antiparallel β-sheet is formed at the dimer interface comprising strands from each of the monomers. The crystal structure suggests a potential mechanism of PrP oligomerization involving three-dimensional domain swapping [Knaus et al., 2001].

Although detailed structural information for PrP^{Sc} is not yet available, studies using low-resolution biophysical techniques (Fourier-transformed infrared spectroscopy, circular dichroism, X-ray fiber diffraction), computational modeling, and analysis of small peptide fragments have led to the

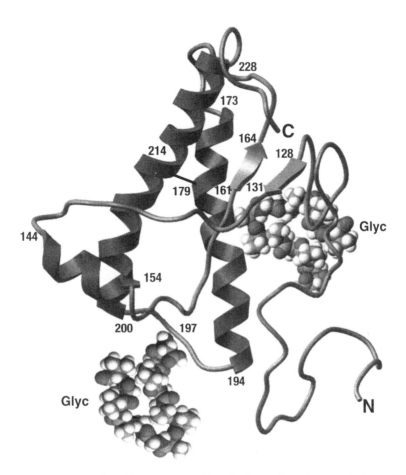

Figure 3.2 Structure of PrP^C. As estimated by NMR studies, the protein has a C-terminal structured domain composed by three α-helices and two short β-strand motifs, and a flexible unordered N-terminal domain.

conclusion that the structural rearrangement produced during PrP^{Sc} formation mostly involves the globular C-terminal domain of the protein [Cohen and Prusiner, 1998; DeMarco and Daggett, 2004]. The current models for PrP^{Sc} structure represent the hydrophobic fragments located on the N-terminal and middle part of the globular domain organized on antiparallel β-sheets [Huang et al., 1996]. These structures become stabilized upon oligomerization with other molecules of PrP^{Sc}.

3.2 Molecular mechanism of PrP^C to PrP^{Sc} conversion

The crucial step in the "protein-only" hypothesis of prion propagation is the conversion of the host PrP^C into the pathological PrP^{Sc}. The notion that endogenous PrP^C is involved in the development of infection is supported by experiments in which the PrP gene was knocked out [Bueler et al., 1993].

These animals were both resistant to prion disease and unable to generate new infectious particles. In addition, it is clear that, during the time between the inoculation with the infectious protein and the appearance of the clinical symptoms, there is a dramatic increase in the amount of PrP^{Sc}. These findings suggest that endogenous PrP^C is converted to the PrP^{Sc} conformation by the action of an infectious form of the PrP molecule. Prion replication is hypothesized to occur when PrP^{Sc} in the infecting inoculum interacts specifically with host PrP^C, catalyzing its conversion to the pathogenic form of the protein. A physical association between the two isoforms during the infectious process is suggested by the primary sequence specificity in prion transmission [Horiuchi et al., 2000] and by the reported *in vitro* generation of PrP^{Sc}-like molecules by mixing purified PrP^C with PrP^{Sc} [Caughey, 2003; Kocisko et al., 1994; Kocisko et al., 1995]. However, the exact mechanism underlying the conversion is not known. From molecular genetic studies as well as from the analysis of the requirements for PrP conversion *in vitro*, it has been postulated that a chaperone-like protein, provisionally called protein X, facilitates PrP^C → PrP^{Sc} conversion (see below). However, the nature of this factor is still unknown.

The precise molecular mechanism of PrP^C → PrP^{Sc} conversion is not well understood. At least two alternative models have been proposed (Figure 3.3), namely:

1. The "template-assisted conversion" model [Cohen and Prusiner, 1998] (Figure 3.3A), which postulates that a physical interaction between the isoforms is a precondition to the structural changes that underlie the pathogenic conversion of PrP^C. In this model, PrP^{Sc} is thermodynamically more stable than PrP^C but is kinetically inaccessible [Cohen and Prusiner, 1998; Harrison et al., 1999]. PrP^C exists in equilibrium with a transient conformational intermediate (named PrP*), which after interaction with a cellular chaperone (protein X) is able to make a heterodimer with PrP^{Sc}. Spontaneously, this heterodimer is converted into a PrP^{Sc} homodimer consisting of the old and newly formed PrP^{Sc} molecules. The homodimer can dissociate to form two templates that are able to induce the conversion, thus generating an exponential growth of PrP^{Sc} concentration [Cohen and Prusiner, 1998].

2. The "nucleation/polymerization" model (Figure 3.3B), which proposes the coexistence of PrP^C and PrP^{Sc} in a thermodynamic equilibrium in solution. PrP^{Sc} monomer is unstable and becomes stabilized upon aggregation with other PrP^{Sc} molecules [Caughey, 2001; Caughey and Lansbury, 2003]. PrP^{Sc} aggregates promote the conversion of PrP^C by binding to the monomeric PrP^{Sc} and displacing the equilibrium toward the formation of the pathological conformer. In this model, the infectious agent is a multimeric, highly ordered aggregate of PrP^{Sc}, and the slow step is the formation of a nucleus that acts as

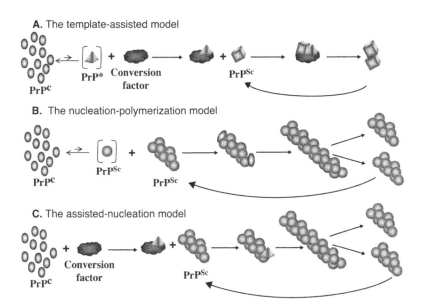

Figure 3.3 Models for PrP^C → PrP^{Sc} conversion. Three different models have been proposed for the mechanism of prion conversion. (A) In the template-assisted conversion model, PrP^C is in equilibrium with an intermediate state (PrP*) that, upon binding with a conversion factor, interacts with PrP^{Sc}, which acts as a template for the conversion. (B) In the nucleation-polymerization model, PrP^C is in equilibrium with an unstable monomeric form of PrP^{Sc}. This form becomes stabilized by oligomerization with the infectious PrP^{Sc} oligomer, which acts as a seed for recruiting, converting, and stabilizing the misfolding of PrP^C. (C) In the assisted-nucleation model, the formation of a partially unfolded intermediate (equivalent to PrP*) is produced upon binding of PrP^C to the conversion factor. The change on the conformation of PrP^C enables PrP* to expose certain hydrophobic regions that are usually buried inside the protein, making possible the interaction with PrP^{Sc}. In this model, PrP^{Sc} is always an oligomer that acts as a seed to convert and stabilize the PrP misfolding, recruiting the newly converted protein into the aggregate. The symbols in brackets refer to a thermodynamically unstable species.

a seed for further stabilization of PrP^{Sc} [Caughey, 2001; Caughey and Lansbury, 2003; Jarrett and Lansbury, Jr., 1993].

A more likely model is, however, in between the template-assisted conversion and the nucleation-polymerization models (Figure 3.3C). In this "assisted-nucleation" model, PrP^{Sc} does not exist as a monomer, but rather a misfolded intermediate (equivalent to PrP*) is formed when PrP^C interact with a cofactor (protein X?). This intermediate exposes the self-recognition motifs required for interaction with PrP^{Sc}, facilitating the binding and incorporation of the yet-unconverted protein into the PrP^{Sc} oligomer. The structural conversion into a β-sheet-rich conformation takes place after binding of PrP* to the PrP^{Sc} polymer, which in this way becomes larger. At some point, the long PrP^{Sc} polymers break into smaller pieces either by a mechanical force or catalyzed by a yet-unknown process. This fragmentation increases the number of effective nuclei, which direct the further conversion of PrP^C.

3.3   Other factors involved in PrP conversion

The idea that others factors are involved in PrP conversion first came from studies with transgenic animals. To explain the results of the transmission of human prions from the brains of CJD (Creutzfeldt-Jakob disease) patients to transgenic mice, Prusiner and colleagues suggested that a macromolecule provisionally named protein X participates in the conversion of PrP^C into PrP^{Sc} [Telling et al., 1995]. In those studies, human prions did not transmit disease to mice coexpressing human and mouse PrP^C, but they did transmit to transgenic mice coexpressing a chimeric mouse-human-mouse PrP^C. Subsequently, transmission of human prions to mice expressing human PrP^C was achieved when the mice were crossed into a PrP null background. These findings were interpreted in terms of mouse PrP^C binding to rodent protein X more avidly than human PrP^C, thus inhibiting the conversion of human PrP^C into PrP^{Sc} [Telling et al., 1995]. Conversely, the binding of the chimeric protein to mouse protein X was similar to that of the mouse prion protein, and hence conversion was not inhibited by the endogenous protein. Later, the same group mapped the region of PrP implicated in the interaction with protein X [Kaneko et al., 1997b] and even produced inhibitors of this interaction.

Using a cell-free conversion assay, we have shown that prion conversion does not occur under our experimental conditions when highly purified PrP^C and PrP^{res} (protease-resistant PrP) are mixed and incubated [Saborio et al., 1999]. Interestingly, the conversion activity was recovered when the bulk of cellular proteins was added back to the sample. This finding provides direct evidence that other factors present in the brain are essential to catalyze prion propagation. As described in Chapter 10, we are using PMCA as a biochemical assay to purify the conversion factor.

3.4   Peptide models used to understand PrP structure and conversion

Investigations with chimeric transgenes showed that PrP^C and PrP^{Sc} are likely to interact within a central domain delimited by codons 96 and 169 [Scott et al., 2000; Telling et al., 1996]. Studies using several synthetic PrP peptides showed that peptides spanning the region 109 to 141 bind to PrP^C and compete with PrP^{Sc} interaction [Chabry et al., 1998; Kaneko et al., 1997a], providing biochemical confirmation for this conclusion. More-detailed biochemical studies should result in a further assessment of the size of the interactive region as well as defining which residues are critical.

The available evidence points to the central fragment of PrP (90–140) as the region mostly involved in PrP conversion. The evidence comes mainly from studies with deletion mutants or short synthetic peptides and includes the following:

- The major component of amyloid fibrils in GSS (Gerstmann-Straussler-Scheinker) brains is a 7- to 11-kDa fragment of PrP that spans residues 58 to 150 [Tagliavini et al., 2001b].
- Residues 1–89 and 141–176 are not required for PrP^C conversion to PrP^{Sc} [Muramoto et al., 1996].
- Deletion of residues 114–121 of PrP results in a protein that is not convertible to PrP^{Sc}, and the modified protein inhibits the conversion of the wild-type PrP^C [Holscher et al., 1998].

Two short synthetic peptides have been extensively used to model different aspects of PrP conformational transition and pathogenicity. Peptides spanning the PrP sequence 109–141 inhibit the interaction between the two PrP isoforms, resulting in prevention of the *in vitro* conversion of PrP^C into PrP^{res} [Chabry et al., 1998]. The same peptide undergoes spontaneous conversion from an initial random coil to a β-sheet secondary structure in a period of a few days [Zhang et al., 1995]. A PrP fragment 109–141 that adopts a β-sheet conformation is resistant to degradation by proteinase K (PK), while the same peptide adopting a random coil structure is rapidly degraded by this protease [Zhang et al., 1995]. Moreover, the PrP fragment 109–141 forms typical amyloid fibrils when incubated under physiological conditions, and the oligomeric β-sheet structure of PrP 109–141 induces rapid death of human neuroblastoma cells [Zhang et al., 1995]. Finally, PrP 109–141 binds tightly to PrP^C and induces a structural rearrangement of the protein, forming a PrP^{Sc}-like isoform [Kaneko et al., 1995]. Kaneko and coworkers reported that PrP^C incubated for 48 hours with a 5000 molar excess of the PrP fragments 90–145 or 109–141 acquires resistance to PK degradation, insolubility, and a higher proportion of β-sheet conformation [Kaneko et al., 1995]. Moreover, the conversion seems to exhibit the known species-barrier characteristics, since Syrian hamster (SHa) PrP^C was transformed into a PK-resistant PrP^{Sc}-like form by a peptide containing the SHa 90–145 sequence but not by a peptide with the corresponding mouse sequence [Kaneko et al., 1995]. However, PK treatment did not result in a shift in the molecular weight, since the electrophoretic mobility of the resistant form was equivalent to that of PrP^C. This result could be interpreted as suggesting that protease resistance might not be due to the generation of PrP^{Sc}, but rather to the formation of a complex between PrP^C and PrP 109–141 oligomers that precludes accessibility to the protease [Kaneko et al., 1997a].

Interestingly, injection of the PrP fragment 90–144 containing the P101L mutation associated with GSS and folded into a β-sheet conformation accelerated the onset of clinical disease in certain transgenic mice [Kaneko et al., 2000]. The same peptide — but not folded in the β-sheet-rich form — did not decrease the incubation time. The effect of the peptide was not observed in wild-type animals or in any other transgenic mice.

The other short peptide that has been extensively used to model the involvement of PrP^{Sc} in TSE is the prion protein fragment spanning the sequence 106–126, corresponding to a putative transmembrane region of PrP^C [Tagliavini et al., 2001a]. PrP 106–126 showed a high intrinsic ability to polymerize *in vitro* and form amyloid fibrils, reminiscent of the scrapie fibrils purified from infected brain [Selvaggini et al., 1993]. In addition, PrP 106–126 is partially resistant to proteinase K and contains a β-sheet-enriched structure [De Gioia et al., 1994; Selvaggini et al., 1993]. The PrP 106–126 peptide has been shown to be cytotoxic *in vitro* via the programmed-cell-death pathway [Forloni et al., 1993]. Recent data suggest that PrP 106–126 also induces apoptotic-mediated cell death *in vivo*; these results were obtained using as a model retinal neurons treated with intravitreous injections of PrP fragments [Ettaiche et al., 2000]. Both *in vitro* and *in vivo*, the toxicity of PrP^{Sc} and PrP 106–126 appears to depend upon neuronal expression of PrP^C and on microglial activation [Bate et al., 2001; Bueler et al., 1993].

An interesting peptide model is the so-called miniprion or PrP 106. Muramoto et al. showed that the minimal sequence required to sustain prion replication in scrapie-infected neuroblastoma cells corresponds to a modified version of PrP lacking residues 23–88 and 141–176 [Muramoto et al., 1996]. PrP 106 has been found to sustain prion replication when expressed in transgenic mice with a PrP knockout genetic background [Supattapone et al., 1999]. Interestingly, recombinant PrP 106 (rPrP 106) spontaneously shows properties similar to those of the PrP 106 extracted from scrapie-infected transgenic mice, such as high β-sheet content, resistance to limited digestion by proteinase K, and high thermodynamic stability [Baskakov et al., 2000]. This fragment has also been produced by chemical synthesis [Bonetto et al., 2002]. Synthetic PrP 106 readily adopted a β-sheet structure, aggregated into branched filamentous structures with ultrastructural and tinctorial properties of amyloid, exhibited a proteinase K-resistant domain spanning residues 134–217, was highly toxic to primary neuronal cultures, and induced a remarkable increase in membrane microviscosity [Bonetto et al., 2002]. These findings suggest that PrP 106 might be an excellent tool for investigating the molecular basis of the conformational conversion of PrP^C into PrP^{Sc} and prion disease pathogenesis.

3.5 Concluding remarks

The prion protein (PrP) can adopt two stable foldings associated with different activities: the normal protein (PrP^C) and the disease-associated isoform (PrP^{Sc}). The evidence indicates that there are no chemical differences between these forms, but they differ in their secondary, tertiary, and quaternary structures. The molecular mechanism for the conversion of PrP^C into PrP^{Sc} has been extensively studied using a variety of model systems and techniques. The conversion apparently involves protein polymerization following a seeding-nucleation process catalyzed by additional brain cofactors. Several critical aspects of this process remain unclear, such as the tridimensional structure of PrP^Sc, the identity of the conversion factors, and the thermodynamic parameters governing the interaction and interconversion of PrP isoforms.

References ***

Baskakov, I.V. et al., Self-assembly of recombinant prion protein of 106 residues, *Biochemistry*, 39, 2792–2804, 2000.

Bate, C., Reid, S., and Williams, A., Killing of prion-damaged neurones by microglia, *Neuroreport*, 12, 2589–2594, 2001.

Bonetto, V. et al., Synthetic miniprion PrP106, *J. Biol. Chem.*, 277, 31327–31334, 2002.

Bueler, H. et al., Mice devoid of PrP are resistant to scrapie, *Cell*, 73, 1339–1347, 1993.

Calzolai, L. et al., Prion protein NMR structures of chickens, turtles, and frogs, *Proc. Natl. Acad. Sci. USA*, 102, 651–655, 2005.

Caughey, B., Interactions between prion protein isoforms: the kiss of death? *Trends Biochem. Sci.*, 26, 235–242, 2001.

**Caughey, B., Prion protein conversions: insight into mechanisms, TSE transmission barriers and strains, *Br. Med. Bull.*, 66, 109–120, 2003.

Caughey, B. and Lansbury, P.T., Protofibrils, pores, fibrils, and neurodegeneration: separating the responsible protein aggregates from the innocent bystanders, *Annu. Rev. Neurosci.*, 26, 267–298, 2003.

*Caughey, B.W. et al., Secondary structure analysis of the scrapie-associated protein PrP 27–30 in water by infrared spectroscopy, *Biochemistry*, 30, 7672–7680, 1991. (Reports one of the first studies in which the secondary structure of misfolded PrP was analyzed.)

Chabry, J., Caughey, B., and Chesebro, B., Specific inhibition of *in vitro* formation of protease-resistant prion protein by synthetic peptides, *J. Biol. Chem.*, 273, 13203–13207, 1998.

Cohen, F.E., Protein misfolding and prion diseases, *J. Mol. Biol.*, 293, 313–320, 1999.

**Cohen, F.E. and Prusiner, S.B., Pathologic conformations of prion proteins, *Annu. Rev. Biochem.*, 67, 793–819, 1998.

De Gioia, L. et al., Conformational polymorphism of the amyloidogenic and neurotoxic peptide homologous to residues 106–126 of the prion protein, *J. Biol. Chem.*, 269, 7859–7862, 1994.

DeMarco, M.L. and Daggett, V., From conversion to aggregation: protofibril formation of the prion protein, *Proc. Natl. Acad. Sci. USA*, 101, 2293–2298, 2004.

Ettaiche, M., Pichot, R., Vincent, J.P., and Chabry, J., In vivo cytotoxicity of the prion protein fragment 106–126, *J. Biol. Chem.*, 275, 36487–36490, 2000.

Forloni, G. et al., Apoptosis mediated neurotoxicity induced by chronic application of beta amyloid fragment 25–35, *Neuroreport*, 4, 523–526, 1993.

Gossert, A.D. et al., Prion protein NMR structures of elk and of mouse/elk hybrids, *Proc. Natl. Acad. Sci. USA*, 102, 646–650, 2005.

Harris, D.A., Trafficking, turnover and membrane topology of PrP, *Br. Med. Bull.*, 66, 71–85, 2003.

---

* Highlight primary articles of outstanding importance and quality, including a short description of the findings.
** Highlights comprehensive review articles related to the topic of this chapter.

Harrison, P.M., Chan, H.S., Prusiner, S.B., and Cohen, F.E., Thermodynamics of model prions and its implications for the problem of prion protein folding, *J. Mol. Biol.*, 286, 593–606, 1999.

Hill, A.F. et al., Molecular classification of sporadic Creutzfeldt-Jakob disease, *Brain*, 126, 1333–1346, 2003.

Holscher, C., Delius, H., and Burkle, A., Overexpression of nonconvertible PrP^C delta114–121 in scrapie-infected mouse neuroblastoma cells leads to trans-dominant inhibition of wild-type PrP(Sc) accumulation, *J. Virol.*, 72, 1153–1159, 1998.

Horiuchi, M., Priola, S.A., Chabry, J., and Caughey, B., Interactions between heterologous forms of prion protein: binding, inhibition of conversion, and species barriers, *Proc. Natl. Acad. Sci. USA*, 97, 5836–5841, 2000.

Huang, Z., Prusiner, S.B., and Cohen, F.E., Scrapie prions: a three-dimensional model of an infectious fragment, *Fold. Des.*, 1, 13–19, 1996.

Jarrett, J.T. and Lansbury, P.T., Jr., Seeding "one-dimensional crystallization" of amyloid: a pathogenic mechanism in Alzheimer's disease and scrapie? *Cell*, 73, 1055–1058, 1993.

Kaneko, K. et al., A synthetic peptide initiates Gerstmann-Straussler-Scheinker (GSS) disease in transgenic mice, *J. Mol. Biol.*, 295, 997–1007, 2000.

Kaneko, K. et al., Prion protein (PrP) synthetic peptides induce cellular PrP to acquire properties of the scrapie isoform, *Proc. Natl. Acad. Sci. USA*, 92, 11160–11164, 1995.

Kaneko, K. et al., Molecular properties of complexes formed between the prion protein and synthetic peptides, *J. Mol. Biol.*, 270, 574–586, 1997a.

Kaneko, K. et al., Evidence for protein X binding to a discontinuous epitope on the cellular prion protein during scrapie prion propagation, *Proc. Natl. Acad. Sci. USA*, 94, 10069–10074, 1997b.

Knaus, K.J. et al., Crystal structure of the human prion protein reveals a mechanism for oligomerization, *Nat. Struct. Biol.*, 8, 770–774, 2001.

*Kocisko, D.A. et al., Cell-free formation of protease-resistant prion protein, *Nature*, 370, 471–474, 1994. (Describes for the first time the *in vitro* conversion of PrP^C into PrP^{Sc}. This system has been very useful in understanding the molecular mechanism of prion replication.)

Kocisko, D.A. et al., Species specificity in the cell-free conversion of prion protein to protease-resistant forms: a model for the scrapie species barrier, *Proc. Natl. Acad. Sci. USA*, 92, 3923–3927, 1995.

Lopez, G.F., Zahn, R., Riek, R., and Wuthrich, K., NMR structure of the bovine prion protein, *Proc. Natl. Acad. Sci. USA*, 97, 8334–8339, 2000.

Lysek, D.A. et al., Prion protein NMR structures of cats, dogs, pigs, and sheep, *Proc. Natl. Acad. Sci. USA*, 102, 640–645, 2005.

*Muramoto, T., Scott, M., Cohen, F.E., and Prusiner, S.B., Recombinant scrapie-like prion protein of 106 amino acids is soluble, *Proc. Natl. Acad. Sci. USA*, 93, 15457–15462, 1996. (Reports the minimal PrP sequence that still sustains prion infection *in vitro* and *in vivo*.)

*Pan, K.M. et al., Conversion of alpha-helices into beta-sheets features in the formation of the scrapie prion proteins, *Proc. Natl. Acad. Sci. USA*, 90, 10962–10966, 1993. (An important study reporting a comparison of secondary structure of PrP^C and PrP^{Sc}.)

Parchi, P. et al., Typing prion isoforms, *Nature*, 386, 232–234, 1997.

*Riek, R. et al., NMR structure of the mouse prion protein domain PrP(121–321), *Nature*, 382, 180–182, 1996. (The first report of the three-dimensional structure of recombinant PrP obtained by NMR.)

*Saborio, G.P. et al., Cell-lysate conversion of prion protein into its protease-resistant isoform suggests the participation of a cellular chaperone, *Biochem. Biophys. Res. Commun.*, 258, 470–475, 1999. (Provides biochemical evidence for the existence of a prion conversion factor.)

Scott, M.R. et al., Transgenic models of prion disease, *Arch. Virol. Suppl.*, 113–124, 2000.

*Selvaggini, C. et al., Molecular characteristics of a protease-resistant, amyloidogenic and neurotoxic peptide homologous to residues 106–126 of the prion protein, *Biochem. Biophys. Res. Commun.*, 194, 1380–1386, 1993. (One of the first studies describing the use of short synthetic peptides to model PrP structural conversions.)

*Stahl, N. et al., Structural studies of the scrapie prion protein using mass spectrometry and amino acid sequencing, *Biochemistry*, 32, 1991–2002, 1993. (A detailed study of the primary structure of PrP^C and PrP^{Sc}, showing the lack of differences between the proteins at this level.)

Supattapone, S. et al., Prion protein of 106 residues creates an artificial transmission barrier for prion replication in transgenic mice, *Cell*, 96, 869–878, 1999.

Tagliavini, F. et al., Studies on peptide fragments of prion proteins, *Adv. Protein Chem.*, 57, 171–201, 2001a.

Tagliavini, F. et al., A 7-kDa prion protein (PrP) fragment, an integral component of the PrP region required for infectivity, is the major amyloid protein in Gerstmann-Straussler-Scheinker disease A117V, *J. Biol. Chem.*, 276, 6009–6015, 2001b.

Telling, G.C. et al., Interactions between wild-type and mutant prion proteins modulate neurodegeneration in transgenic mice, *Genes Dev.*, 10, 1736–1750, 1996.

*Telling, G.C. et al., Prion propagation in mice expressing human and chimeric PrP transgenes implicates the interaction of cellular PrP with another protein, *Cell*, 83, 79–90, 1995. (The first study providing evidence for the existence of a prion conversion factor.)

Zahn, R. et al., NMR solution structure of the human prion protein, *Proc. Natl Acad. Sci. USA*, 97, 145–150, 2000.

Zhang, H. et al., Conformational transitions in peptides containing two putative alpha-helices of the prion protein, *J. Mol. Biol.*, 250, 514–526, 1995.

chapter four

Cell biology, genetic and putative function of the normal prion protein

The evolutionary conservation of the prion protein (PrP) primary structure among mammals, its high level of expression in brain tissue, and its localization in highly specialized membrane domains suggest that the normal cellular protein (PrP^C) may play an important biological function. It is not yet known whether changes in the putative biological activity of PrP^C are implicated in the pathogenesis [Hetz et al., 2003]. This chapter describes the cell biology, biosynthesis, and cellular localization of PrP^C as well as the several putative biological activities associated with it. The contribution of animal models to our understanding of PrP^C biological activity and its participation in TSE (transmissible spongiform encephalopathy) pathogenesis is addressed at the end of the chapter.

4.1 Cellular biology of the normal prion protein

The biosynthetic pathway followed by PrP^C is similar to that of other membrane and secreted proteins; PrP^C is synthesized in the endoplasmic reticulum (ER) and transits the Golgi apparatus through the cell surface. During its biosynthesis in the ER, the N-terminal signal peptide is removed, N-linked oligosaccharide chains are added, the disulfide bond is formed, and the GPI (glycosylphosphatidylinositol) anchor is attached following cleavage of the C-terminal signal peptide [Harris, 2003]. The N-linked oligosaccharide chains added in the ER are of the high-mannose type and are sensitive to digestion by endoglycosidase H. These are subsequently modified in the Golgi apparatus to yield complex-type chains that contain sialic acid and are resistant to endoglycosidase H.

Most of mature PrP^C is located in specialized membrane structures known as lipid rafts, rich in spingolipids and cholesterol [Taraboulos et al., 1992] (Figure 4.1). Rafts have been experimentally defined as detergent-resistant

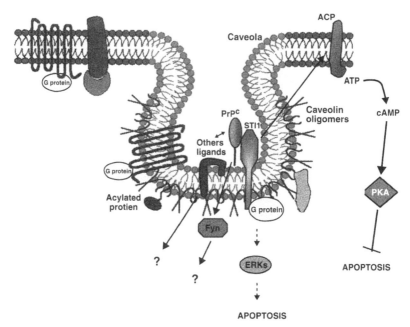

Figure 4.1 (See color insert after p. 114.) Signal transduction through the prion protein. Under physiological conditions, PrP^C is located in membrane-specific domains known as lipid rafts or caveola. These membrane structures are rich in receptors and proteins involved in signaling. Interaction between PrP^C and one of its putative ligands (STI1) leads to the potential activation of several signaling pathways: (i) activation of a putative G-protein (Gp), which in turn can activate an adenilate cyclase (ACP), leading to the formation of cAMP and the consequent activation of PKA, and (ii) activation of the ERKs pathway promoting neuronal death. In addition, activation of the tyrosine kinase Fyn by PrP^C via a not-yet-understood mechanism may lead to a signaling pathway with unknown consequences. Finally, interaction of PrP^C with other ligands (laminin receptor, bcl2, neural cell-adhesion molecules, etc.) may also trigger some signaling events.

membrane domains and functionally defined by the presence of many important cellular receptors, tyrosine kinases, and most of the GPI-anchored proteins [Simons and Ehehalt, 2002; Simons and Toomre, 2000]. Part of the PrP^C molecules constitutively cycle between the plasma membrane and an endocytic compartment [Harris, 2003]. Kinetic studies have shown that PrP^C molecules cycle through the cell with a transit time of around 60 min, and during each passage, 1 to 5% of the molecules undergo proteolytic cleavage near residue 110 [Harris, 1999; Harris, 2003]. Internalized PrP^C colocalizes with the endosomal markers transferrin and FM4–64.

An aspect that remains to be elucidated is the subcellular localization of PrP most relevant for TSE pathogenesis [Harris, 2003; Lasmezas, 2003]. Until recently, it was thought that membrane GPI-anchored PrP either exposed in the cell surface or in the endocytic pathway was the location most relevant for PrP^C biological function as well as the site of conversion. However, recent evidence suggests that cytosolic or transmembrane PrP molecules exist and may play a role in disease pathogenesis. In some inherited cases of prion diseases, it has been reported that the predominant form of PrP detectable in the brain is not PrP^{Sc}, but rather a transmembrane form of the prion protein, named ^{CTM}PrP [Hegde et al., 1998; Hegde et al., 1999]. The role of ^{CTM}PrP in TSE pathogenesis remains unclear, and future research should help in clarifying whether this form is an irrelevant by-product or a triggering factor in neurodegeneration. Recently, it was reported that cells overexpressing PrP undergo cytosolic accumulation of a protease-resistant PrP^{Sc}-like form due to retrograde transport though the ER when proteasome activity is impaired [Ma et al., 2002; Ma and Lindquist, 2002]. The mechanism involved in neuronal cell death induced by proteasome dysfunction remains to be determined.

4.2   A signaling role for the prion protein?

The prion protein is constitutively and highly expressed on neurons and is evolutionarily conserved. The location of PrP^C in lipid rafts, a membrane fraction specialized in signaling, provided a first hint of a potential role of PrP^C in signal transduction. Support for this idea came from antibody-mediated cross-linking experiments showing a caveolin-1-dependent coupling of PrP^C to Fyn, a member of the Src family of tyrosine kinases [Mouillet-Richard et al., 2000] (Figure 4.1). Although these results indicate that PrP^C may function as a signaling molecule, the authors did not report any of the expected morphological responses downstream of Fyn. Other signaling proteins have also been proposed to bind PrP^C. Using a yeast two-hybrid approach, neuronal phosphoprotein synapsin Ib, the adapter protein Grb2, and the still uncharacterized prion interactor protein Pint1 were identified as potential interacting proteins [Spielhaupter and Schatzl, 2001]. Moreover, the *in vivo* interaction of these proteins with PrP^C was supported by coimmunoprecipitation, although no indication of the physiological relevance of these interactions is yet available.

Evidence to support the hypothesis that PrP^C plays a role in neuroprotective signaling comes from complementary hydropathy analysis, which identified the stress-inducible protein 1 (STI1) as a potential PrP^C-interacting protein in the plasma membrane [Zanata et al., 2002] (Figure 4.1). STI1 is a heat-shock protein, first described in a complex with the Hsp70 and Hsp90 chaperone family of proteins [Lassle et al., 1997]. Using organotypical retinal explants, it was shown that the interaction between PrP^C and STI1 protects neurons from anisomycin-induced apoptosis [Zanata et al., 2002]. This phenomenon was not observed in tissues derived from PrP knockout mice. A detailed analysis of the signaling pathways involved in PrP^C-mediated neuroprotection was performed using a peptide ligand identified by complementary hydropathy [Chiarini et al., 2002]. PrP^C interaction with this peptide induced activation of the cyclic adenosine monophosphate (cAMP)-dependent protein kinase A (PKA) and ERK (extracellular signal-regulated kinases) pathways (Figure 4.1). Activation of the PKA pathway was involved in the inhibition of anisomycin-induced apoptosis, whereas activation of the ERK pathway promoted apoptosis. These findings suggest that the balance between pro- and antiapoptotic signals may determine the final physiological consequence of PrP^C-mediated signal transduction. Interestingly, the peptide ligand used to activate PrP^C signaling was shown to interact with the PrP fragment 113–128. Based on these findings, it was proposed that PrP^C interacts with an endogenous ligand (possibly STI1) and constitutively transduces a survival signal through the PKA pathway. In addition, during the development of prion diseases, the conformational changes associated with PrP conversion might result in decreased affinity for the ligand, eliminating the PrP^C-mediated neuroprotective effect. Additional support for a neuroprotective role of PrP^C comes from studies with neuronal primary cultures showing that *Prnp$^{-/-}$* cells are more sensitive than wild-type cells to apoptotic stimuli, such as serum deprivation [Kuwahara et al., 1999].

Despite the evidence in favor of a neuroprotective signaling role for PrP^C, it is not clear how a GPI-anchored protein can initiate a signaling cascade, since activation of adenyl cyclase is dependent on the activity of G-proteins located in the inner side of the plasma membrane. Thus, the possible mechanism of PrP^C-mediated signaling remains speculative, and more research is required to clarify this key issue.

4.3   *PrP^C ligands indicate a potential role in apoptosis*

The search for PrP-interacting proteins has been pursued in an attempt to understand the function of PrP^C. In addition to the proteins described above, other candidate ligands for PrP^C include Bcl-2, Hsp60, Bip, Nrf-2, apolipoprotein A1, neural cell-adhesion molecules, heparin, laminin, and laminin receptor (reviewed in [Gauczynski et al., 2001; Martins et al., 2002]).

PrP fusion proteins interact with Bcl-2, as shown by yeast two-hybrid experiments and by coimmunoprecipitation [Kurschner and Morgan, 1996]. A functional relationship between PrP^C and the Bcl-2 antiapoptosis family is suggested by experiments showing that Bcl-2 overexpression can attenuate the increased susceptibility to serum-deprivation-induced apoptosis of PrP null neurons *in vitro* [Kuwahara et al., 1999]. A link between PrP and the Bcl-2 family is also suggested by the similarity between the highly conserved octapeptide repeats of PrP and the Bcl-2 homology domain 2 (BH2) common to the Bcl-2 protein family. This domain is crucial for the antiapoptotic function of Bcl-2 and its interaction with the proapoptotic protein Bax, suggesting that PrP might be a member of the antiapoptotic family. Support for this hypothesis comes from *in vitro* experiments showing that PrP^C can protect human neurons against Bax-induced apoptosis [Bounhar et al., 2001]. This neuroprotective activity was not observed with mutant proteins bearing deletions of the octapeptide repeats or with some C-terminal PrP mutations associated with inherited prion diseases [Bounhar et al., 2001]. These results are consistent with the observation on transgenic mice expressing mutant

PrP devoid of the octapeptide repeats, which show alterations in the incubation time and histopathology during infections with scrapie [Flechsig et al., 2000]. The major problem with the hypothesis of PrP^C being an antiapoptotic protein like Bcl2 is that it has not been shown to be present in cytoplasm, mitochondria, or on the outer surface of endoplasmic reticulum, where this family of proteins normally exerts its activity.

4.4 A putative role of PrP^C in copper metabolism

Another widely studied putative function of PrP^C is on the binding and metabolism of copper [Brown, 2001]. The link between copper and TSEs comes from the old studies showing a prion-like spongiform degeneration of the brain by treatment of animals with the copper chelator cuprizone [Pattison and Jebbett, 1973]. The octapeptide repeat of PrP^C is able to bind copper within the physiological concentration range [Brown et al., 1997; Kramer et al., 2001], suggesting that PrP^C may have a role in normal brain metabolism of copper. PrP^C overexpression was shown to increase copper uptake into cells and enhance copper incorporation into superoxide dismutase. Furthermore, the prion protein itself was shown to have a weak superoxide dismutase activity *in vitro* [Brown et al., 1999], which is dependent on copper binding to the octapeptide repeats. Interaction between PrP^C and either PrP^{Sc} or the PrP 106–126 peptide inhibits the superoxide dismutase activity, suggesting that PrP misfolding, with a concomitant loss of enzyme function, may contribute to neuronal damage [Brown, 2001].

In vivo experiments have shown that *Prnp* knockout mice have lower levels of copper in their brains and elevated copper in serum compared with wild-type mice [Brown et al., 1997]. However, another report found no differences in the copper levels in transgenic animal brains expressing different quantities of PrP^C [Waggoner et al., 2000]. On the other hand, it has been shown that, in the brains of scrapie-infected mice prior to the onset of clinical symptoms, there is a significant change in the levels of copper [Thackray et al., 2002; Wong et al., 2001b]. Analysis of purified PrP from the brains of scrapie-infected mice also showed a reduction in copper binding to the protein and a proportional decrease in antioxidant activity between 30 and 60 days postinoculation [Thackray et al., 2002]. In addition, in humans affected by sCJD (sporadic Creutzfeldt-Jakob disease), there is a decrease of up to 50% of copper in the brain [Wong et al., 2001a]. The conclusion is that (a) PrP^C may be an important brain membrane copper-binding protein that participates in copper homeostasis and (b) PrP misfolding may result in abnormalities of metal levels, thus leading to neuronal damage [Brown, 2001].

PrP^C is constantly endocytosed, passing through an early endocytic compartment before being restored to the plasma membrane [Harris, 1999; Taraboulos et al., 1992]. Copper ions have been reported to rapidly and reversibly stimulate the internalization of PrP^C [Pauly and Harris, 1998]. It has been proposed that copper binding increases the affinity of PrP^C for an endocytic receptor, leading to PrP^C and copper internalization. Therefore, PrP^C could serve as a recycling receptor for uptake of copper ions from the extracellular milieu. Copper interaction with PrP may also modulate the conversion of the normal protein into the pathological isoform. Indeed, it has been shown that the binding of Cu^{2+} causes a change in both the secondary and tertiary structure of PrP, leading to the formation of a protease-resistant form [Qin et al., 2000]. Copper binding increases the stability of the β-sheet and thermodynamically promotes the shift from α-helical to the β-sheet associated with PrP^{Sc}.

4.5   *PrP knockout animals and doppel*

Several PrP-null animals were generated with the aim to understand the possible biological function of PrP^C. However differences in the way gene disruption was achieved have led to conflicting conclusions [Behrens and Aguzzi, 2002; Rossi et al., 2001]. A consistent finding in all mice devoid of PrP^C is their resistance to scrapie, and their inability to propagate infectivity. The first two null mice lines generated, referred to as *Prnp*$^{0/0}$ Zurich I and *Prnp*$^{-/-}$ Edinburgh, were viable and had no obvious neurological problems [Bueler et al., 1992; Manson et al., 1994], although some minor electrophysiological and circadian rhythm defects were observed. In contrast, three subsequently generated knockout lines — termed *Prnp*$^{-/-}$ Nagasaki, Rmc0, and *Prnp*$^{0/0}$ Zurich II — developed cerebellar Purkinje cell degeneration and demyelination of peripheral nerves, causing late-onset ataxia [Moore et al., 1999; Rossi et al., 2001; Sakaguchi et al., 1996]. This phenotype was abolished by introduction of a PrP transgene, suggesting that a loss of PrP^C activity was responsible for neurodegeneration in these animals. The discrepancy in the phenotype of the different PrP-null animals was resolved by a detailed comparison of the way in which disruption of the PrP gene was achieved [Moore et al., 1999; Rossi et al., 2001]. This analysis led to the discovery by Westaway and coworkers of a gene located downstream of PrP, termed *Prnd*, which encodes a 179-residue protein named doppel (Dpl) [Moore et al., 1999]. Dpl has sequential and structural homology to the C-terminal two thirds of PrP^C, but it lacks the unordered N-terminal region. Like PrP^C, it is attached to the cell surface through a GPI anchor [Nicholson et al., 2002]. *Prnd* is expressed at high levels in testes and some peripheral organs, but in very low amounts in the adult brain [Tranulis et al., 2001]. Mice carrying a homozygous disruption of the *Prnd* gene have a normal embryonic and postnatal development, but males are sterile [Behrens et al., 2002].

In the PrP^C-null animals that developed ataxia, an artifactual intergenic splicing places *Prnd* under the control of the PrP promoter, leading to high levels of Dpl expression in the brain. The onset of ataxia and Purkinje cell loss is directly dependent on increased levels of Dpl in the central nervous system (CNS) [Behrens and Aguzzi, 2002; Moore et al., 2001; Rossi et al., 2001], and thus neurodegeneration in these animals is apparently due to a neurotoxic activity of Dpl, rather than to the loss of a putative PrP^C

neuroprotective activity. These findings led to the suggestion that some of the TSE clinical symptoms such as ataxia might be due to increased or unscheduled expression and activity of Dpl in brain during the disease. This possibility was investigated by grafting *Prnd*-null embryonic stem cells into wild-type mice brain and studying TSE pathogenesis in the resulting strain. The results indicate that Dpl deficiency does not prevent prion pathogenesis [Behrens et al., 2001]. This conclusion is supported by a recent study showing that TSE infection does not alter the levels of *Prnd* in the CNS and that Dpl expression in brain has no influence on the incubation period, spongiform degeneration, or PrP^{Sc} deposition in TSE-infected mice [Tuzi et al., 2002]. Thus, it seems that expression of Dpl in brain does not contribute to TSE pathogenesis. However, it is not possible to rule out a role for Dpl in peripheral prion replication and neuroinvasion, especially considering that *Prnd* is expressed in spleen, an organ known to be implicated in peripheral prion pathogenesis.

The molecular mechanism by which Dpl overexpression in the brain leads to cerebellar degeneration is a subject of current investigation. In the Nagazaki and Zurich II PrP-null mice, Dpl neurotoxicity was rescued by introduction of a *Prnp* transgene, suggesting that PrP^C can antagonize Dpl activity and that the absence of PrP^C is a necessary prerequisite for Dpl to induce cell death [Behrens and Aguzzi, 2002; Moore et al., 2001; Rossi et al., 2001]. Interestingly, a previous study showed that expressing a truncated PrP molecule lacking the fragment 32–134 or 32–121 of PrP (but not smaller fragments) in the Zurich I mouse background results in ataxia and degeneration of the cerebellar granule cell layer [Shmerling et al., 1998]. Moreover, this phenotype could be abolished by the reintroduction of a single wild-type PrP allele. Because the truncated transgene in these experiments lacks the region of PrP^C that is also absent in Dpl, it is tempting to speculate that Dpl and truncated PrP induce neurodegeneration by a similar mechanism.

Three different hypotheses have been proposed to explain the mechanism by which PrP^C antagonizes Dpl neurotoxicity [Behrens and Aguzzi, 2002; Hetz et al., 2003]. First, PrP^C and Dpl may compete for interaction with an unknown receptor or ligand that promotes neuronal survival. While PrP^C has a higher affinity and acts as an agonist, Dpl behaves as an antagonist because its binding is weak or incomplete. Only in the absence of PrP^C is Dpl able to bind the ligand, resulting in neurotoxicity, because an inactive complex is formed that is either unable to elicit a survival signal or triggers apoptosis. In the presence of PrP^C, the higher affinity of the ligand for its natural partner prevents the interaction with Dpl and the consequent neurotoxicity. In the absence of both PrP^C and Dpl in the brain, binding sites on the ligand remain unoccupied, in which case it is necessary to postulate that either the ligand itself has an intrinsic ability to signal survival, or another hypothetical protein, previously dubbed π, is able to bind and exert PrP^C-like function [Shmerling et al., 1998]. This would explain why PrP knockout mice show no phenotype [Shmerling et al., 1998]. According to this model, PrP^C and Dpl act through the same receptor/ligand, but with opposite effects.

A second possibility is that PrP^C normally has a neuroprotective activity, while Dpl expressed in the brain generates a proapoptotic stimulus. In the absence of PrP^C, Dpl leads to cell loss and ataxia, whereas in the presence of the neuroprotective activity of PrP^C, the negative effect of Dpl is suppressed. In this model, the binding partners of PrP^C and Dpl are postulated to be different and the effects generated are independent. It supports the idea that the physiological function of PrP^C is related to neuronal survival, which, given the mild phenotype of the PrP-null mutation, is perhaps manifested only in the presence of injury or an apoptotic stimulus. This model explains less well why the phenotype associated with expression of truncated PrP^C resembles that resulting from expression of Dpl.

A third model proposes that Dpl expression in the brain results in a gain of a neurotoxic activity through an intrinsic amyloidogenic tendency to oligomerize and that when PrP^C is reintroduced into the brain it becomes part of the oligomer, making it inactive or unstable. In this hypothesis, the amyloid-like properties of Dpl are manifested only when it is expressed in its nonphysiological location, and PrP^C acts as a trans-suppressor of protein aggregation.

Drawing some of these ideas together, we have proposed an alternative model in which PrP^C has two different functional domains: a neuroprotective domain located within the N-terminal fragment of PrP, and a neurotoxic domain hidden in the C-terminal region of the protein [Hetz et al., 2003] (Figure 4.2). A balance between the two opposite activities of PrP^C might explain both the biological function of the protein and its implication in TSE pathogenesis. Under physiological conditions, the neuroprotective activity prevails because the neurotoxic sequence is hidden inside the protein, whereas during TSE pathogenesis, the protein conformational changes expose the neurotoxic domain and perhaps also inhibit the neuroprotective function, leading to neuronal apoptosis. Because Dpl and truncated PrP contain only the C-terminal domain, their expression in the absence of PrP^C leads to neurodegeneration, which is minimized by the neuroprotective effect of PrP^C when reintroduced into the brain. In the absence of both Dpl and PrP^C, the neuroprotective and neurotoxic activities of the protein are balanced, resulting in no obvious phenotypic changes other than the hypersensitivity of the animals to apoptotic stimuli and oxidative stress. This model is consistent with the evidence showing multiple activities of PrP associated with different regions of the molecule (Figure 4.2), which could potentially activate counteracting signaling pathways. Furthermore, it does not impose the need to postulate unknown ligands or compensatory mechanisms. A prediction of this hypothesis is that Dpl toxicity and perhaps TSE neurodegeneration might be decreased by overexpression of the N-terminal fragment of PrP.

Taken together, the data from knockout mice suggest that animals lacking PrP^C develop normally and do not exhibit major phenotypic changes that might give a clear indication of the physiological function of PrP^C. This may be due to the activation of compensatory feedback mechanisms during

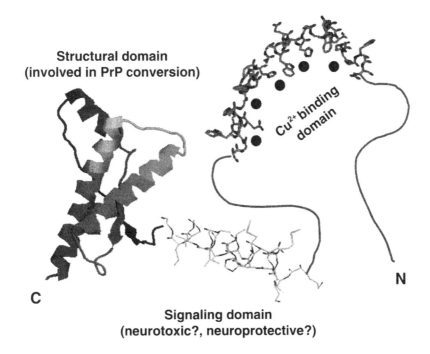

**Structural domain
(involved in PrP conversion)**

Cu²⁺-binding
domain

N

C

**Signaling domain
(neurotoxic?, neuroprotective?)**

Figure 4.2 (See color insert after p. 114.) PrP^C domains involved in diverse cellular functions. The model of PrP^C is based on the known NMR structure of the protein, which indicates the protein regions associated with PrP^C activity. In purple is shown the amino-terminal region in which the five octapeptide repeats are located. This domain has been associated with copper binding, SOD activity, protection from oxidative stress, and Bax-induced apoptosis. In red is shown the neurotoxic domain from amino acids 106–126, which has been involved in the interaction with STI1 protein and the initiation of PKA and ERK signaling. The carboxy-terminal globular domain is implicated in the conversion of PrP^C into PrP^Sc.

development or because the alterations are too subtle to produce neurodegeneration or other overt clinical signs.

To assess the possibility of compensatory mechanisms during development, a conditional postnatal *Prnp*^-/- knockout mouse was recently generated [Mallucci et al., 2002]. These mice remained healthy with no evidence of neurodegeneration or other histopathological changes for up to 15 months postknockout. However, neurophysiological evaluation showed significant reduction of afterhyperpolarization potentials in hippocampal CA1 cells, suggesting a direct role for PrP in the modulation of neuronal excitability. The authors concluded that acute depletion of PrP does not affect neuronal survival in this model, ruling out loss of PrP function as a cause of TSE [Mallucci et al., 2002]. Despite this, Chiarini and coworkers speculated that short-term compensation in the postnatal knockout mice could attenuate the consequence of acute deletion of PrP^C [Chiarini et al., 2002]. In addition, the possibility that PrP^C deletion leads to more-subtle changes not resulting in extensive neurodegeneration is supported by a recent study of a more detailed biochemical analysis of PrP knockout mice [Brown et al., 2002]. This study showed the presence of several markers of oxidative stress, and the animals were more sensitive to neurological damage. No evident tissue degeneration was observed, although cells derived from these animals were hypersensitive to oxidative stress. The biochemical changes of these mice include increased levels of nuclear factor NF-kB and Mn dismutase, decreased levels of p53, altered melatonin levels, and increased expression of apoptosis-related genes Bax and Bcl-2 and phospho-ERKs [Brown et al., 2002]. These results are consistent with the proposal that PrP^C may participate in a neurotrophic signaling pathway, but that the lack of the protein does not lead to damage unless an external toxic stimulus is present. According to this scenario, we could speculate that TSE arises from a combination of a toxic activity of PrP^{Sc} exacerbated by the loss of a neurotrophic activity of PrP^C.

4.6   Concluding remarks

A more complete understanding of PrP^C biological function and its participation in the maintenance of neuronal activity is key to understanding whether a loss of function of PrP^C may be implicated in the pathogenesis of prion disorders. Resolving this question is crucial for the development of therapeutic approaches for TSE, because if reduced PrP^C physiological function is involved in disease progression, then strategies directed toward maintaining or replacing this activity could provide novel opportunities for treatment of the prion-related diseases. Several diverse functions have been attributed to PrP^C, and although the physiological relevance of any of them remains unclear, it seems possible that PrP^C is a multifunction protein, with different and separate domains serving different purposes (Figure 4.2). The interpretation of the current data is that, although PrP^C may be implicated in neuronal function and survival and copper metabolism, it is unlikely that TSE neurodegeneration is explained exclusively by a loss of PrP^C activity.

References* **

Behrens, A. et al., Normal neurogenesis and scrapie pathogenesis in neural grafts lacking the prion protein homologue doppel, EMBO Rep., 2, 347–352, 2001.

Behrens, A. et al., Absence of the prion protein homologue doppel causes male sterility, EMBO J., 21, 3652–3658, 2002.

Behrens, A. and Aguzzi, A., Small is not beautiful: Antagonizing functions for the prion protein PrP(C) and its homologue Dpl, Trends Neurosci., 25, 150–154, 2002.

---

* Highlights primary articles of outstanding importance and quality, including a short description of the findings.
** Highlights comprehensive review articles related to the topic of this chapter.

Bounhar, Y. et al., Prion protein protects human neurons against Bax-mediated apoptosis, *J. Biol. Chem.*, 276, 39145–39149, 2001.

Brown, D.R., Copper and prion disease, *Brain Res. Bull.*, 55, 165–173, 2001.

Brown, D.R., Nicholas, R.S., and Canevari, L., Lack of prion protein expression results in a neuronal phenotype sensitive to stress, *J. Neurosci. Res.*, 67, 211–224, 2002.

*Brown, D.R. et al., The cellular prion protein binds copper in vivo, *Nature*, 390, 684–687, 1997. (An important study showing that the normal prion protein binds copper under physiological conditions *in vivo*, providing strong evidence for a putative role in copper metabolism.)

Brown, D.R. et al., Normal prion protein has an activity like that of superoxide dismutase, *Biochem. J.*, 344 (pt. 1), 1–5, 1999.

*Bueler, H. et al., Normal development and behaviour of mice lacking the neuronal cell-surface PrP protein, *Nature*, 356, 577–582, 1992. (A seminal paper demonstrating that the lack of a normal prion protein did not lead to any abnormal phenotype in animals.)

*Chiarini, L.B. et al., Cellular prion protein transduces neuroprotective signals, *EMBO J.*, 21, 3317–3326, 2002. (Reports some compelling evidence in cell culture for a potential role of PrP^C in signal transduction.)

Flechsig, E. et al., Prion protein devoid of the octapeptide repeat region restores susceptibility to scrapie in PrP knockout mice, *Neuron*, 27, 399–408, 2000.

Gauczynski, S. et al., Interaction of prion proteins with cell surface receptors, molecular chaperones, and other molecules, *Adv. Protein Chem.*, 57, 229–272, 2001.

**Harris, D.A., Cellular biology of prion diseases, *Clin. Microbiol. Rev.*, 12, 429–444, 1999.

**Harris, D.A., Trafficking, turnover and membrane topology of PrP, *Br. Med. Bull.*, 66, 71–85, 2003.

*Hegde, R.S. et al., A transmembrane form of the prion protein in neurodegenerative disease, *Science*, 279, 827–834, 1998. (This intriguing study describes a new cellular location for PrP^C. Although its relevance for the disease is not clear, this finding changed our views on the trafficking and metabolism of PrP^C.)

Hegde, R.S. et al., Transmissible and genetic prion diseases share a common pathway of neurodegeneration, *Nature*, 402, 822–826, 1999.

**Hetz, C., Maundrell, K., and Soto, C., Is loss of function of the prion protein the cause of prion disorders? *Trends Mol. Med.*, 9, 237–243, 2003.

Kramer, M.L. et al., Prion protein binds copper within the physiological concentration range, *J. Biol. Chem.*, 276, 16711–16719, 2001.

Kurschner, C. and Morgan, J.I., Analysis of interaction sites in homo- and heteromeric complexes containing Bcl-2 family members and the cellular prion protein, *Brain Res. Mol. Brain Res.*, 37, 249–258, 1996.

*Kuwahara, C. et al., Prions prevent neuronal cell-line death, *Nature*, 400, 225–226, 1999. (An intriguing study reporting an increase of cellular survival in cells expressing PrP^C vs. knockout cells.)

**Lasmezas, C.I., Putative functions of PrP(C), *Br. Med. Bull.*, 66, 61–70, 2003.

Lassle, M. et al., Stress-inducible, murine protein mSTI1: characterization of binding domains for heat shock proteins and in vitro phosphorylation by different kinases, *J. Biol. Chem.*, 272, 1876–1884, 1997.

*Ma, J. and Lindquist, S., Conversion of PrP to a self-perpetuating PrPSc-like conformation in the cytosol, *Science*, 298, 1785–1788, 2002. (Another intriguing experiment showing that stressing cells led PrP^C to convert into a protease-resistant form that self-propagates in the cytoplasm.)

Ma, J., Wollmann, R., and Lindquist, S., Neurotoxicity and neurodegeneration when PrP accumulates in the cytosol, *Science*, 298, 1781–1785, 2002.

*Mallucci, G.R. et al., Post-natal knockout of prion protein alters hippocampal CA1 properties, but does not result in neurodegeneration, *EMBO J.*, 21, 202–210, 2002. (A very important study that clarifies the real effect of knocking down PrP expression.)

Manson, J.C. et al., PrP gene dosage determines the timing but not the final intensity or distribution of lesions in scrapie pathology, *Neurodegeneration*, 3, 331–340, 1994.

**Martins, V.R. et al., Cellular prion protein: on the road for functions, *FEBS Lett.*, 512, 25–28, 2002.

*Moore, R.C. et al., Ataxia in prion protein (PrP)-deficient mice is associated with upregulation of the novel PrP-like protein doppel, *J. Mol. Biol.*, 292, 797–817, 1999. (Reports the discovery of the PrP homolog doppel and clarifies the discrepancy on the phenotype of diverse PrP knockout mice.)

Moore, R.C. et al., Doppel-induced cerebellar degeneration in transgenic mice, *Proc. Natl. Acad. Sci. USA*, 98, 15288–15293, 2001.

Mouillet-Richard, S. et al., Signal transduction through prion protein, *Science*, 289, 1925–1928, 2000.

Nicholson, E.M. et al., Differences between the prion protein and its homolog doppel: a partially structured state with implications for scrapie formation, *J. Mol. Biol.*, 316, 807–815, 2002.

Pattison, I.H. and Jebbett, J.N., Clinical and histological recovery from the scrapie-like spongiform encephalopathy produced in mice by feeding them with cuprizone, *J. Pathol.*, 109, 245–250, 1973.

Pauly, P.C. and Harris, D.A., Copper stimulates endocytosis of the prion protein, *J. Biol. Chem.*, 273, 33107–33110, 1998.

Qin, K. et al., Copper(II)-induced conformational changes and protease resistance in recombinant and cellular PrP: effect of protein age and deamidation, *J. Biol. Chem.*, 275, 19121–19131, 2000.

Rossi, D. et al., Onset of ataxia and Purkinje cell loss in PrP null mice inversely correlated with Dpl level in brain, *EMBO J.*, 20, 694–702, 2001.

Sakaguchi, S. et al., Loss of cerebellar Purkinje cells in aged mice homozygous for a disrupted PrP gene, *Nature*, 380, 528–531, 1996.

Shmerling, D. et al., Expression of amino-terminally truncated PrP in the mouse leading to ataxia and specific cerebellar lesions, *Cell*, 93, 203–214, 1998.

Simons, K. and Ehehalt, R., Cholesterol, lipid rafts, and disease, *J. Clin. Invest.*, 110, 597–603, 2002.

Simons, K. and Toomre, D., Lipid rafts and signal transduction, *Nat. Rev. Mol. Cell Biol.*, 1, 31–39, 2000.

Spielhaupter, C. and Schatzl, H.M., PrP^C directly interacts with proteins involved in signaling pathways, *J. Biol. Chem.*, 276, 44604–44612, 2001.

Taraboulos, A. et al., Synthesis and trafficking of prion proteins in cultured cells, *Mol. Biol. Cell*, 3, 851–863, 1992.

Thackray, A.M. et al., Metal imbalance and compromised antioxidant function are early changes in prion disease, *Biochem. J.*, 362, 253–258, 2002.

Tranulis, M.A. et al., The PrP-like protein doppel gene in sheep and cattle: cDNA sequence and expression, *Mamm. Genome*, 12, 376–379, 2001.

Tuzi, N.L. et al., Expression of doppel in the CNS of mice does not modulate transmissible spongiform encephalopathy disease, *J. Gen. Virol.*, 83, 705–711, 2002.

Waggoner, D.J. et al., Brain copper content and cuproenzyme activity do not vary with prion protein expression level, *J. Biol. Chem.*, 275, 7455–7458, 2000.

Wong, B.S. et al., Aberrant metal binding by prion protein in human prion disease, *J. Neurochem.*, 78, 1400–1408, 2001a.

Wong, B.S. et al., Increased levels of oxidative stress markers detected in the brains of mice devoid of prion protein, *J. Neurochem.*, 76, 565–572, 2001b.

Zanata, S.M. et al., Stress-inducible protein 1 is a cell surface ligand for cellular prion that triggers neuroprotection, *EMBO J.*, 21, 3307–3316, 2002.

chapter five

Prion strains, species barriers, and multiple conformations of the prion protein

It is well established that the prion infectious agent, like conventional micro-organisms, exhibits strain variation and species barrier. Diverse prion strains have been identified in mice, hamsters, sheep, cattle, and humans. Strains differ on disease characteristics, including incubation period, neuropathology, and clinical symptoms. In addition, prions from one animal species can infect only a limited number of other animal species. Understanding how a single protein can provide the diversity to sustain the variety of strains and the species-barrier phenomenon has been a challenge for the prion hypothesis [Prusiner, 1998; Soto and Castilla, 2004]. This chapter reviews the characteristics of prion strains and the species-barrier process and discusses the evidence indicating that the prion protein (PrP) structure enciphers these phenomena.

5.1 Prion strains

Studies of scrapie in sheep and goats and later in mice demonstrated reproducible variations in disease phenotype with the passage of prions in genetically inbred hosts [Pattison and Millson, 1961; Kimberlin, 1976]. The distinct varieties or isolates of prions were called "strains," in analogy to viral strains. Strains can be defined as subspecies of an infectious agent capable of maintaining specific phenotypic profiles when passed to an experimental animal. Strains can differ in several characteristics, including incubation time, brain-lesion profile, clinical manifestation of the disease, and biochemical properties of PrP [Bruce, 2003].

A method of distinguishing prion strains was developed by Bruce, Manson, and colleagues. This method relies on measuring the elapsed time from infection to the onset of clinical disease in several inbred mouse strains, and then assessing the clinical signs and the distribution of tissue damage in various brain regions [Bruce and Fraser, 1991; Carp et al., 1994; Bruce, 2003]. Transmissible spongiform encephalopathy (TSE) in ordinary nontransgenic mice is characterized by long asymptomatic incubation periods lasting between about 4 months and the full life span of the mouse. Following this presymptomatic phase, progressive neurodegeneration and clinical signs appear over a period of a few weeks. Despite the length of the interval between exposure to infection and the clinical phase, if all experimental conditions are kept constant, the incubation period is remarkably predictable [Bruce, 2003]. However, different TSE strains tested in the same mouse strain give markedly different incubation periods.

The incubation period is profoundly dependent on the genetic makeup of the host [Bruce et al., 1991]. In mice, two alleles of the PrP gene have been recognized (designated a and b), encoding proteins that differ by two amino acids at codons 108 (leucine or phenylalanine) and 189 (threonine or valine) [Westaway et al., 1987]. When mice are infected with a single TSE strain, the PrP genotype can make a difference of hundreds of days to the incubation period. Interestingly, the effects of genotypic differences on disease progression were identified by Dickinson and coworkers in the late 1960s, long before the prion protein was discovered [Dickinson et al., 1968]. Initially, the gene was called *Sinc* (for scrapie incubation), with the two *Sinc* alleles s7 and p7 corresponding to the a and b alleles of the PrP gene.

TSE strains interact with the PrP gene in a complex way, with each strain producing a characteristic and highly reproducible pattern of incubation time in the three possible PrP mouse genotypes (the two homozygotes and the heterozygote F_1 cross). For example, the mouse background showing the shortest incubation period could be PrP-a for some TSE strains and PrP-b for other strains. The incubation period in the F_1 cross (PrP-ab) lies sometimes between those of PrP-a and PrP-b mice and sometimes beyond the longer of the two, but is never shorter than both.

TSE strains also show dramatic differences in the neuropathological damage with respect to the type, severity, and distribution of the changes [Fraser, 1993]. In routine histological sections, the vacuolation profile is different in distinct strains, sometimes concentrated in a specific brain area, whereas in other strains it is mostly spread throughout the brain. This is the basis of a semiquantitative method of strain discrimination in which the severity of vacuolation is scored from coded sections in nine gray-matter and three white-matter brain areas to construct a "lesion profile" that is characteristic for each combination of TSE strain and mouse genotype [Fraser and Dickinson, 1968; Bruce et al., 1991]. TSE strains may also differ in the extent, brain region, and compactness of the accumulation of pathological prion proteins (PrP^{Sc}). With most TSE strains, pathological deposits of PrP can readily be demonstrated in the brain in the form of diffuse deposits in areas of vacuolation and, more focally, as amyloid plaques [Bruce et al., 1989]. Some TSE strains target PrP pathology precisely to particular groups of neurons, leaving the surrounding brain regions unaffected. Other strains produce a more generalized pathology, albeit with a preference for particular brain areas. Some TSE strains produce many amyloid plaques, while others produce few or none. These observations suggest a fundamental difference in PrP^{Sc} replication and accumulation between TSE strains, which is most likely the reason why different prion strains are characterized by distinct clinical symptoms, depending on the brain region damaged.

In humans, up to seven different strains of sCJD (sporadic Creutzfeldt-Jakob disease) have been identified based on the clinical features, prion protein genotype at codon 129, distribution of histopathological damage, and Western blot profile of PrP^{Sc} [Parchi et al., 1999] (Table 5.1). Interestingly, in human PrP, the polymorphism at position 129, where a valine or a methione could be present, determines and controls prion strains as well as disease incidence and onset.

In addition to the interesting scientific problem that the concept of prion strains raises, there are important implications for public health. For example, it is now clear that there are two strains of bovine spongiform encephalopathy (BSE), one of which has been well documented to infect humans, producing vCJD (variant Creutzfeldt-Jakob disease) [Prusiner, 1997; Collinge, 1999; Bruce, 2000]. The other, a recently described strain that is pathologically characterized by the presence of PrP-immunopositive amyloid plaques (as opposed to the lack of amyloid deposition in typical BSE cases) and by a different pattern of regional distribution and topology of brain PrP^{Sc} accumulation [Casalone et al., 2004]. In addition, Western blot analysis makes it possible to biochemically distinguish the PrP^{Sc} isoforms associated with both BSE strains. Strikingly, the molecular signature of this previously undescribed bovine PrP^{Sc} was similar to that encountered in a distinct subtype of sCJD, suggesting that transmission of disease from cattle to humans might be more widespread than originally thought [Casalone et al., 2004]. Most disturbingly, there is grave concern that one or both of these BSE strains might have infected sheep, where they could produce a disease hardly distinguishable from scrapie (a long-known prion disease of sheep, probably innocuous to humans) [Foster et al., 2001]. But if its threatening strain-specific properties are maintained across the species barrier, sheep BSE may be a threat to human health.

5.2 Species barrier

One of the characteristics of the agent responsible for prion diseases is its ability to infect some species and not others. This phenomenon is known as the species barrier. Even between close species where the barrier is rather small, the species barrier is manifested as a prolongation of the time it takes for animals to develop the clinical disease when injected with another species's infectious material [Hill and Collinge, 2004]. There are two distinct events in trans-species transmission of TSE. The first is the primary recruitment events where the trans-species PrP^{Sc} is directly involved in conversion

Table 5.1 Clinical, Histological, and Biochemical Features of Different Human Prion Strains

sCJD Type	Major Clinical Symptoms [a]	Brain Area Damaged	Pattern of PrP^{Sc} Deposition	Proportion of PrP^{Sc} Glycoforms [b]
MM1	Myoclonus, dementia, rigidity, visual problems	Cortex, thalamus, parasubiculum, neostriatum	Mostly diffuse synaptic (one-third cases perivacuolar)	2.4 : 4.5 : 3.1
MV1	Myoclonus, dementia, ataxia, rigidity	Cortex, thalamus, parasubiculum, neostriatum	Mostly diffuse synaptic (one-quarter cases perivacuolar)	2.4 : 4.5 : 3.1
VV1	Dementia, aphasia, apraxia, myoclonus, rigidity	Cortex, thalamus, parasubiculum, neostriatum	Mostly diffuse synaptic	2.1 : 4.4 : 3.5
MM2-C	Dementia, aphasia, rigidity, myoclonus	Cortex, thalamus, parasubiculum, neostriatum	Mostly perivacuolar	3.3 : 4.1 : 2.6
MM2-T	Dementia, ataxia, dysarthria, insomnia, psychiatric problems	Thalamus, medulla, parasubiculum	Low PrP^{Sc} staining either synaptic or perivacuolar	2.5 : 4.1 : 3.4
MV2	Ataxia, dementia, rigidity, myoclonus	Parasubiculum, neostriatum, thalamus, cortex	Extensive amyloid plaques in cerebellum and PrP^{Sc} staining	2.8 : 4.1 : 3.1
VV2	Ataxia, dementia, myoclonus	Cerebellum, neostriatum, thalamus, parasubiculum	Cerebellar amyloid plaques and diffuse synaptic deposits	3.3 : 4.1 : 2.6

[a] The clinical symptoms listed correspond to those observed in at least 50% of the patients. The list follows the order of frequency observed.

[b] Proportion of the three PrP^{Sc} glycosylation forms (di- : mono- : unglycosylated) as analyzed by Western blot after proteinase K digestion.

Source: Data in this table represents a summary of the findings on the postmortem evaluation of 300 subjects affected by sCJD. From Parchi, P. et al., *Ann. Neurol.*, 46, 224–233, 1999. With permission.

of the host's normal cellular protein (PrP^C), followed by secondary self-recruitment whereby newly formed PrP^{Sc} converts the host's normal PrP^C [Kellershohn and Laurent, 1998]. Serial passages determine that the initial PrP^{Sc} disappears, though not necessarily without a trace because it imparts the newly formed PrP^{Sc} with a new strain type [Hill and Collinge, 2004].

The species barrier has been studied extensively in the lab using experimental rodents and, in particular, transgenic mice. On the first transmission of a natural TSE to mice, the incubation period is usually very long, and there may be survivors [Bruce, 2003]. In subsequent serial mouse-to-mouse transmissions, the incubation period shortens and stabilizes, usually after three to four passages. The neuropathological profile also stabilizes in the course of these first few passages. Thereafter, the incubation period and neuropathological characteristics are stable indefinitely upon further mouse-to-mouse passages [Bruce, 2003].

Compelling evidence indicates that the species barrier is largely controlled by the sequence of PrP. Although prion proteins are evolutionarily conserved in mammals, there are some differences that may affect the interaction of PrP^{Sc} from one species with the PrP^{C} from another (Figure 5.1). Unfortunately, we cannot yet predict the degree of a species barrier simply by comparing the prion proteins from two species; the barrier has to be measured by experimental studies in animals. Obviously, one cannot assess species barriers in humans, as this would require deliberate infection. One method of doing this is to use genetically modified mice that produce human prion proteins. Another possibility is to evaluate the *in vitro* conversion of human PrP^{C} challenged by PrP^{Sc} from different species [Raymond et al., 1997].

The importance of PrP sequence in species barrier was elegantly demonstrated in studies of prion transmission between hamsters and mice. Mice are normally resistant to infection from the widely used 263K strain of

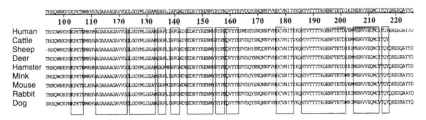

Figure 5.1 Sequence alignment of PrP from different mammals. The sequence chosen for alignment was the PrP region 94–226, which corresponds to the globular domain of the protein associated with the conformational change featuring the conversion of PrP^{C} into PrP^{Sc}. The most common mammalian species naturally affected by TSE (human, cattle, sheep, deer, and mink), the experimental rodents used to model the disease (hamster and mouse), and two species known not to be susceptible to propagate prions (rabbit and dog) were chosen for the alignment. The sequence on top represents the consensus among the different species. The sequence inside the frame represents conserved regions among all species studied (no amino acid differences).

hamster prion, but they become susceptible to this strain following trans-genic introduction of the hamster PrP gene [Scott et al., 1989]. In addition, disruption of the endogenous mouse gene and introduction of a hamster transgene renders mice highly susceptible to hamster prions and resistant to mouse prions [Prusiner et al., 1993]. These findings suggest that the species barrier is controlled by a direct interaction between infecting PrP^{Sc} and endogenous PrP^C, in which effective interaction is inhibited by differences in the PrP sequence. Between mouse and hamster, such a barrier must be mediated by one or more of the 16 differences (out of 254 residues) in the PrP sequence (Figure 5.1). Scott and colleagues studied the amino acid dif-ferences most important for species barrier using transgenic mice expressing chimeric PrP genes derived from hamster (Ha) and mouse (M) PrP genes [Scott et al., 1993]. One line of transgenic mice, designated MHa2M PrP, contains a gene with five amino acid substitutions encoded by hamster PrP, while another construct, designated MHaM2 PrP, has two substitutions. MHa2M PrP transgenic mice were susceptible to both hamster and mouse prions, whereas three lines expressing MHaM2 PrP were resistant to Syrian hamster prions [Scott et al., 1993].

From other studies with transgenic mice, it has been found that, in addition to the PrP sequences, two other factors may contribute to the species barrier. The first is the strain of prion, which as discussed earlier seems to be enciphered in the conformation of PrP^{Sc} and works together with the PrP sequence to determine the tertiary structure of nascent PrP^{Sc}. The second is the species specificity of protein X, a conversion factor implicated in the transformation of PrP^C into PrP^{Sc} (see Chapter 4). Protein X was postulated to explain the somehow puzzling results on the transmission of human prions to transgenic mice [Telling et al., 1994; Telling et al., 1995]. Mice expressing both mouse and human PrP were resistant to human prions, whereas those expressing only human PrP were susceptible. These results suggest that mouse PrP^C inhibited transmission of human prions. Strikingly, mice expressing both mouse PrP and a chimeric gene containing pieces of the mouse and human sequence (designated MHu2M PrP) were susceptible to human prions, and mice expressing the chimeric gene alone were only slightly more susceptible [Telling et al., 1994; Telling et al., 1995]. These findings indicate that mouse PrP^C has only a minimal effect on the formation of chimeric MHu2M PrP^{Sc}. The interpretation of these results suggested that mouse PrP^C binds to mouse protein X with a higher affinity than human PrP^C and that mouse PrP^C had little effect on the formation of PrP^{Sc} from MHu2M because mouse PrP and MHu2M PrP share the same amino acid sequence at the COOH terminus, which was postulated as the binding site for protein X [Kaneko et al., 1997].

5.3 *Multiple conformations of PrP^{Sc}*

The existence of the prion strains and the phenomenon of the species barrier has been one of the strongest arguments to oppose the prion hypothesis (see

Chapter 2) [Chesebro, 1998; Soto and Castilla, 2004]. If prions would contain nucleic acids, it should be easy to explain strains as the result of genetic polymorphisms. On the other hand, if prions consist solely of a misfolded protein, the phenotypic traits of each strain must in some way be encrypted within the pathological prion protein. Because several distinct prion strains can be maintained in the same animal background, the primary structure of PrP cannot contribute to strain-specific properties. To resolve this conundrum, it was suggested that strains may be specified by distinct infectious conformations of PrP^{Sc}, each of which is capable of imparting its particular three-dimensional imprint to host PrP^C [Prusiner, 1998; Cohen and Prusiner, 1998]. The first hint that strain specificity may reside within the structure of PrP^{Sc} came from the work of Marsh, Bessen, and coworkers, who identified differences in the electrophoretic mobility patterns of PrP^{Sc} derived from two strains of mink prions (Figure 5.2A) [Bessen and Marsh, 1994]. Distinct patterns of PrP^{Sc} also exist in human Creutzfeldt-Jakob disease (Figure 5.2B). The groups of Collinge and Gambetti have shown that these patterns reflect unique combinations of glycoforms that are maintained upon transmission to mice [Collinge et al., 1996; Parchi et al., 1996] (Figure 5.2B). Furthermore, it was found that vCJD produces the same glycoform pattern as BSE, adding considerable momentum to the hypothesis that vCJD prions originated from BSE prions [Collinge et al., 1996].

More direct support for the structural differences among PrP^{Sc} associated with different strains came from several reports analyzing the secondary structure of PrP^{Sc} from diverse strains [Caughey et al., 1998; Safar et al., 1998; Aucouturier et al., 1999]. Structural studies of PrP^{Sc} purified from diverse hamsters, mouse, and human prion strains have shown reproducible differences in the content of β-sheet as measured by Fourier-transformed infrared spectroscopy [Caughey et al., 1998; Safar et al., 1998; Aucouturier et al., 1999]. In addition, Safar and colleagues have developed an immunological assay, termed conformational dependent immunoassay (CDI), capable of distinguishing PrP^{Sc} from diverse strains based on the differential recognition by antibodies of the native and denatured states of the protein [Safar et al., 1998]. Recently Surewicz's group has shown that aggregates of recombinant PrP from different species differ not only on their secondary structure, but also in the morphological and structural properties of the fibrillar aggregates as studied by atomic force microscopy [Jones and Surewicz, 2005].

5.4 Concluding remarks

One of the most intriguing aspects for the protein-based infectious agents associated with TSE is their ability to exist as different strains that faithfully propagate different disease features. These strains usually rise when a prion from one species is adapted into a new species by breaking the so-called species barrier. Recent findings suggest that the phenomena of prion strains and species barrier are enciphered and controlled by the sequence and structure of PrP^{Sc}. Although there is compelling evidence to support this

A

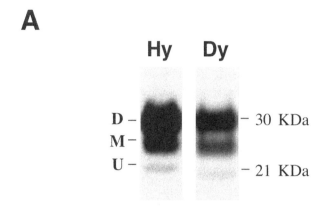

B

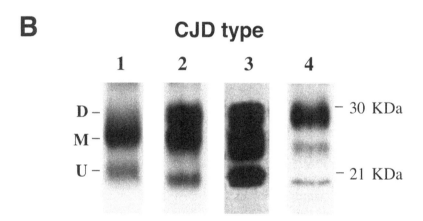

Figure 5.2 Western blot profile for PrP^{Sc} associated with different strains. (A) Brains from two hamster strains termed hyper (Hy) and drowsy (Dy) were analyzed by Western blot after proteinase K digestion. (B) Four human strains associated with sCJD (type 1 and 2), iatrogenic CJD (type 3), and vCJD (type 4) show significant differences on the PrP^{Sc} profile both in terms of the proportion of the different glycoforms and the migration of the unglycosylated band. D, M, and U represent di-, mono-, and unglycosylated bands, respectively.

hypothesis, there are still many open questions regarding the mechanism by which strains replicate and the tightness of the species barrier. These questions are not only important to understand the unprecedented biology of prions, they also have important implications for public health. With the recent development of strategies to generate infectious prions *in vitro* [Legname et al., 2004; Castilla et al., 2005], it is likely that the coming years will bring a better understanding of the molecular basis of prion strains and the species barrier.

References* **

Aucouturier, P., Kascsak, R.J., Frangione, B., and Wisniewski, T., Biochemical and conformational variability of human prion strains in sporadic Creutzfeldt-Jakob disease, *Neurosci. Lett.*, 274, 33–36, 1999.

*Bessen, R.A. and Marsh, R.F., Distinct PrP properties suggest the molecular basis of strain variation in transmissible mink encephalopathy, *J. Virol.*, 68, 7859–7868, 1994. (One of the first reports providing a molecular explanation for the prion strains.)

Bruce, M.E., "New variant" Creutzfeldt-Jakob disease and bovine spongiform encephalopathy, *Nature Med.*, 6, 258–259, 2000.

**Bruce, M.E., TSE strain variation, *Br. Med. Bull.*, 66, 99–108, 2003.

**Bruce, M.E. and Fraser, H., Scrapie strain variation and its implications, *Curr. Top. Microbiol. Immunol.*, 172, 125–138, 1991.

Bruce, M.E., McBride, P.A., and Farquhar, C.F., Precise targeting of the pathology of the sialoglycoprotein, PrP, and vacuolar degeneration in mouse scrapie, *Neurosci. Lett.*, 102, 1–6, 1989.

Bruce, M.E., McConnell, I., Fraser, H., and Dickinson, A.G., The disease characteristics of different strains of scrapie in Sinc congenic mouse lines: implications for the nature of the agent and host control of pathogenesis, *J. Gen. Virol.*, 72, 595–603, 1991.

Carp, R.I., Ye, X., Kascsak, R.J., and Rubenstein, R., The nature of the scrapie agent. Biological characteristics of scrapie in different scrapie strain-host combinations, *Ann. N. Y. Acad. Sci.*, 724, 221–234, 1994.

*Casalone, C. et al., Identification of a second bovine amyloidotic spongiform encephalopathy: molecular similarities with sporadic Creutzfeldt-Jakob disease, *Proc. Natl. Acad. Sci. USA*, 101, 3065–3070, 2004. (A very important study providing evidence for a second strain of BSE.)

Castilla, J., Saá, P., Hetz, C., and Soto, C., *In vitro* generation of infectious scrapie prions, *Cell*, 121, 195–206, 2005.

*Caughey, B., Raymond, G.J., and Bessen, R.A., Strain-dependent differences in beta-sheet conformations of abnormal prion protein, *J. Biol. Chem.*, 273, 32230–32235, 1998. (Reports differences in secondary structure of PrP^{Sc} associated with different strains.)

Chesebro, B., BSE and prions: uncertainties about the agent, *Science*, 279, 42–43, 1998.

Cohen, F.E. and Prusiner, S.B., Pathologic conformations of prion proteins, *Annu. Rev. Biochem.*, 67, 793–819, 1998.

Collinge, J., Variant Creutzfeldt-Jakob disease, *Lancet*, 354, 317–323, 1999.

*Collinge, J., Sidle, K.C., Meads, J., Ironside, J., and Hill, A.F., Molecular analysis of prion strain variation and the aetiology of "new variant" CJD, *Nature*, 383, 685–690, 1996. (Reports a biochemical method for strain typing *in vitro* based on the Western blot profile of PrP^{Sc} and provides important molecular evidence for the BSE origin of vCJD.)

---

* Highlights primary articles of outstanding importance and quality, including a short description of the findings.
** Highlights comprehensive review articles similar to the topic of this chapter.

'Dickinson, A.G., Meikle, V.M., and Fraser, H., Identification of a gene which controls the incubation period of some strains of scrapie agent in mice, *J. Comp Pathol.*, 78, 293–299, 1968. (An early and pioneering report of a genetic difference responsible for TSE strains. Fifteen years later, this gene was found to be PrP.)

Foster, J.D. et al., Clinical signs, histopathology and genetics of experimental transmission of BSE and natural scrapie to sheep and goats, *Vet. Rec.*, 148, 165–171, 2001.

Fraser, H., Diversity in the neuropathology of scrapie-like diseases in animals, *Br. Med. Bull.*, 49, 792–809, 1993.

Fraser, H. and Dickinson, A.G., The sequential development of the brain lesion of scrapie in three strains of mice, *J. Comp Pathol.*, 78, 301–311, 1968.

"Hill, A.F. and Collinge, J., Prion strains and species barriers, *Contrib. Microbiol.*, 11, 33–49, 2004.

'Jones, E.M. and Surewicz, W.K., Fibril conformation as the basis of species- and strain-dependent seeding specificity of mammalian prion amyloids, *Cell*, 121, 63–72, 2005. (A recent study showing that the basis for prion strain and species barrier may be differences in the morphological and structural features of PrP aggregates.)

Kaneko, K. et al., Evidence for protein X binding to a discontinuous epitope on the cellular prion protein during scrapie prion propagation, *Proc. Natl. Acad. Sci. USA*, 94, 10069–10074, 1997.

Kellershohn, N. and Laurent, M., Species barrier in prion diseases: a kinetic interpretation based on the conformational adaptation of the prion protein, *Biochem. J.*, 334 (Pt. 3), 539–545, 1998.

Kimberlin, R.H., Experimental scrapie in mouse: review of an important model disease, *Science Progress*, 63, 461–481, 1976.

Legname, G. et al., Synthetic mammalian prions, *Science*, 305, 673–676, 2004.

'Parchi, P. et al., Molecular basis of phenotypic variability in sporadic Creutzfeldt-Jakob disease, *Ann. Neurol.*, 39, 767–778, 1996. (Reports a biochemical method for strain typing *in vitro* based on the Western blot profile of PrPSc.)

'Parchi, P. et al., Classification of sporadic Creutzfeldt-Jakob disease based on molecular and phenotypic analysis of 300 subjects, *Ann. Neurol.*, 46, 224–233, 1999. (An extensive and comprehensive study of the clinical, histological, and biochemical characteristics of different human sCJD strains.)

Pattison, I.H. and Millson, G.C., Scrapie produced experimentally in goats with special reference to the clinical syndrome, *J. Comp Pathol.*, 71, 101–109, 1961.

Prusiner, S.B., Prion diseases and the BSE crisis, *Science*, 278, 245–251, 1997.

Prusiner, S.B., Prions, *Proc. Natl. Acad. Sci. USA*, 95, 13363–13383, 1998.

Prusiner, S.B. et al., Ablation of the prion protein (PrP) gene in mice prevents scrapie and facilitates production of anti-PrP antibodies, *Proc. Natl. Acad. Sci. USA*, 90, 10608–10612, 1993.

'Raymond, G.J. et al., Molecular assessment of the potential transmissibilities of BSE and scrapie to humans, *Nature*, 388, 285–288, 1997. (An interesting study reporting the evaluation of the strength of species barrier using the cell-free conversion assay.)

'Safar, J. et al., Eight prion strains have PrP(Sc) molecules with different conformations, *Nature Med.*, 4, 1157–1165, 1998. (Reports differences in secondary structure of PrPSc associated with different strains and describes an immunological assay to detect those differences.)

Scott, M. et al., Transgenic mice expressing hamster prion protein produce spe-cies-specific scrapie infectivity and amyloid plaques, *Cell*, 59, 847–857, 1989.

Scott, M. et al., Propagation of prions with artificial properties in transgenic mice expressing chimeric PrP genes, *Cell*, 73, 979–988, 1993.

Soto, C. and Castilla, J., The controversial protein-only hypothesis of prion propaga-tion, *Nature Med.*, 10, S63–S67, 2004.

Telling, G.C. et al., Transmission of Creutzfeldt-Jakob disease from humans to trans-genic mice expressing chimeric human-mouse prion protein, *Proc. Natl. Acad. Sci. USA*, 91, 9936–9940, 1994.

Telling, G.C. et al., Prion propagation in mice expressing human and chimeric PrP transgenes implicates the interaction of cellular PrP with another protein, *Cell*, 83, 79–90, 1995.

Westaway, D. et al., Distinct prion proteins in short and long scrapie incubation period mice, *Cell*, 51, 651–662, 1987.

chapter six

From the mouth to the brain

Transmissible spongiform encephalopathies (TSEs) are diseases affecting primarily the brain. The fastest and most efficient method for inducing spongiform encephalopathy in laboratory animals is intracerebral inoculation of brain homogenate. However, prion diseases can also be initiated by feeding, by intravenous and intraperitoneal injection, as well as from the eye by conjunctival instillation, corneal grafts, and intraocular injection. The peripheral routes are probably the most relevant in the natural spread of the disease.

TSEs typically exhibit a very long latency period between the time of infection and the clinical manifestation. Evidence suggests that during this time, prions undergo a first replication phase in peripheral lymphoid tissues before moving to the brain [Aguzzi, 2003]. During this process, little or no damage occurs to the brain, and thus it should be possible to design therapeutic strategies aimed to block the entry of prions, preventing neurodegeneration. This chapter discusses the state-of-the-art in the knowledge of the pathways and mechanisms implicated in the transport of prion infectious agent from the mouth to the brain (Figure 6.1).

6.1 Prions in the gastrointestinal tract

One of the puzzling findings in prion transmission is that the infectious agent composed exclusively of a pathological prion protein (PrP^{Sc}) can resist the low pH and high protease concentrations of the gut and penetrate the intestinal barrier to get into the blood, lymphoid organs, peripheral nerves, and finally the brain (Figure 6.1). The high resistance of prions to procedures that normally destroy proteins is probably due to the unique biochemical and structural properties of PrP^{Sc}. As discussed in Chapter 3, PrP^{Sc} is highly resistant to protease digestion, extreme temperatures, and chemical denaturation.

Upon oral challenge, an early rise in prion infectivity can be observed in the distal ileum of infected individuals [Wells et al., 1994], where Peyer's patches acquire strong immunoreactivity for the prion protein. Evidence indicates that Peyer's patches may represent a portal of entry for orally

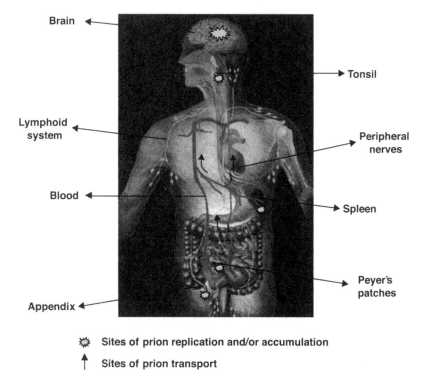

Brain

Tonsil

Lymphoid
system

Peripheral
nerves

Blood

Spleen

Peyer's
patches

Appendix

☼ Sites of prion replication and/or accumulation

↑ Sites of prion transport

Figure 6.1 (See color insert after p. 114.) Schematic representation of the pathways implicated in the transport and replication of prions. Orally ingested prions are intestinally absorbed mainly at the level of Peyer's patches and transported to the blood and lymphoid fluids. After a peripheral replication in spleen, appendix, tonsil, and other lymphoid tissues, prions are transported to the brain mainly via peripheral nerves. A direct penetration to the brain across the blood–brain barrier is also possible.

administered prions on their transport from the mouth to the brain [Press et al., 2004]. However, the mechanism by which prions pass across the intestinal barrier is not known, and only a limited number of studies have been done [Ghosh, 2004]. *In vitro* studies using cellular models of intestinal permeability have shown that membranous epithelial cells (M cells) (Figure 6.2) play an important role in transepithelial transport of prions [Heppner et al., 2001], as has also been shown for the passage of enteric pathogens. Administration of scrapie prions to the apical compartment of cocultures consisting of Caco-2 cells, B-lymphoblastoid cells, and M cells allowed transport of infectivity toward the basolateral side [Heppner et al., 2001]. In contrast, there was hardly any prion transport in Caco-2 cultures without M cells. These findings indicate that M-cell differentiation is necessary for active transepithelial prion transport *in vitro* [Aguzzi et al., 2003]. Uptake of foreign antigens by M cells has been shown to be followed by rapid transcytosis directly to the intraepithelial pocket, where macrophages,

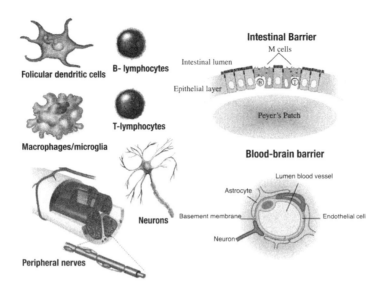

Figure 6.2 (See color insert after p. 114.) Cells implicated in prion diseases. Diverse cell types have been implicated in the transport of prions (B- and T-lymphocytes, peripheral nerves, M cells), in the replication of the infectious agent (neurons, follicular dendritic cells), and in the clearance of prions (macrophages, microglia). Also, to reach the brain, the infectious agent has to cross at least two tight barriers, the intestinal and the blood–brain barrier.

dendritic cells, and lymphocytes are located (Figure 6.2), establishing a direct connection to the immune system [Neutra et al., 1996].

6.2 The immune-system connection

Lymphoid organs have long been known to be involved in the early steps of prion diseases [Aguzzi et al., 2003; Aguzzi, 2003; Aucouturier et al., 2000]. In particular, the spleen and lymph nodes have been demonstrated to be the first sites of PrP^{Sc} replication after infection by peripheral routes and also to be significantly affected following intracerebral challenge. In humans, abundant PrP^{Sc} was demonstrated in the germinal centers of the tonsils and the appendix in patients affected by variant Creutzfeldt-Jakob disease (vCJD) [Wadsworth et al., 2001].

Compelling evidence has been accumulated over the years in support of a major role of lymphoid organs in prion peripheral propagation and transport [Aguzzi et al., 2003; Aguzzi, 2003; Aucouturier et al., 2000]. Several indirect findings suggest that the lymphoid system influences the course of TSE. For example, (a) susceptibility to prion infection was shown to correlate with the maturation of the immune system [Outram et al., 1973]; (b) corticosteroids reduce the susceptibility to scrapie [Outram et al., 1974]; and

(c) mitogenic stimulation of lymphoid cells enhances the susceptibility to scrapie [Dickinson et al., 1978]. More direct evidence was the report that splenectomy significantly delayed the onset of scrapie in mice infected by the peripheral route, while it did not affect the incubation period after intracerebral infection [Fraser and Dickinson, 1970]. In a similar way, studies on mice with severe immunodeficiencies showed that these animals were partially resistant to scrapie after intraperitoneal or subcutaneous inoculation [Klein et al., 1997]. Replication of prions in lymphoid organs was also demonstrated after intracerebral inoculation, raising the question of a role for lymphoid organs not only on prion transport, but also in the disease pathogenesis [Kimberlin and Walker, 1979].

The nature of the immune-cell types that support this initial step of infection still remain unclear. There are controversial results regarding the respective roles of lymphocytes and follicular dendritic cells (FDC) in the replication of prions and their transport to the nervous system [Aucouturier and Carnaud, 2002]. Although mice that lack B cells do not develop neurological signs of scrapie infection after peripheral challenge [Klein et al., 1997], they may nevertheless accumulate PrP^{Sc} and infectivity in the brain [Frigg et al., 1999]; thus, the clinical expression of the disease might partly rely upon certain aspects of the immunological status. After fractionation of spleen cells by density-gradient centrifugation, the highest infectivity was found in lymphocytes a few weeks after inoculation [Kuroda et al., 1983]. The normal prion protein is consistently expressed, albeit at relatively low levels compared with the brain, in circulating lymphocytes [Cashman et al., 1990]. Interestingly, in the experiments reported by Klein and colleagues, the presence of B cells was essential for appearance of the disease in the brain [Klein et al., 1997], but the expression of PrP was not [Klein et al., 1998]. On the other hand, T-cell deficiency brought about by ablation of the T-cell receptor α-chain did not affect prion pathogenesis [Brandner et al., 1999]. Lymphocytes alone could not account for the peripheral prion pathogenesis, because adoptive transfer of *Prnp*$^{+/+}$ bone marrow to *Prnp*$^{0/0}$ recipient mice was not sufficient to restore infectibility of *Prnp*-expressing brain grafts [Blattler et al., 1997]. These findings indicate that the presence of B cells — but not expression of the cellular prion protein by these cells — is indispensable for pathogenesis upon intraperitoneal infection in the mouse scrapie model, indicating that prion replication in these cells was not necessary [Aguzzi et al., 2003].

Another point that is unclear is the nature of the cells that replicate and accumulate prions in lymphoid organs [Aucouturier and Carnaud, 2002]. Although it is known that PrP^{Sc} exists in a variety of lymphoid organs, it is not yet clear which cells accumulate the misfolded protein. It is important to note that the experiments showing that lymphocytes play a role in prion transport to the brain do not necessarily indicate that these cells are the primary repository of prions in lymphoid tissues [Aguzzi et al., 2003]. Actually, the results from the experiments of bone marrow grafting have suggested that most splenic prion infectivity may reside in a "stromal" fraction.

Therefore, lymphocytes may be important for trafficking prions within lymphoid organs, but they do not appear to represent the major place for prion replication in the periphery [Aguzzi et al., 2003]. FDCs are a good candidate for prion replication, since they express large amounts of cellular prion protein, and PrP accumulations tend to colocalize with FDCs in light- and electron-microscope analyses of prion-infected spleens [Jeffrey et al., 2000]. However, the presence of the scrapie agent in FDCs could reflect their potent antigen-capturing function [Banchereau and Steinman, 1998]. A key role for FDCs in PrP^{Sc} replication is strongly supported by experiments showing that the expression of PrP^C by FDC, but not by hematopoietic lineages, is required for scrapie susceptibility after peripheral inoculation [Brown et al., 1999]. However, these findings conflict with some of the previous reports showing the importance of lymphocytes for prion propagation. It has been suggested that previous results showing a major role for B-lymphocytes in prion neuroinvasion may have simply reflected the importance of these cells for maintaining maturation of FDCs [Aguzzi, 2003; Aucouturier and Carnaud, 2002].

Elegant studies from Aguzzi's group have demonstrated the importance of FDCs for prion pathogenesis. Gene-deletion experiments in mice have shown that signaling by the tumor necrosis factor (TNF) and lymphotoxins is required for FDC development. By altering the maturation of FDCs using a lymphotoxin-α/β receptor-immunoglobulin fusion protein, these authors showed that lack of mature FDCs led to a significant delay or disappearance of the disease after intraperitoneal administration of the prion infectious agent [Montrasio et al., 2000]. These data indicate that FDCs are essential for deposition of PrP^{Sc} and generation of infectivity in the spleen, and suggest that FDCs participate in the process of neuroinvasion.

6.3 From the lymphoid organs to the brain: peripheral nerves or blood–brain barrier?

Although the role of lymphoid organs on the peripheral prion propagation is very well established, it is much less clear how the infectious agent goes from there to the brain. PrP^{Sc} could leave lymphoid organs and reach the central nervous system via nerve fibers or the blood stream (Figure 6.1). Both mechanisms may be hypothesized on the basis of existing data, and these possibilities are not mutually exclusive. Detection of the scrapie agent in organs such as lung and intestine soon after it appeared in lymphoid organs suggested that it may spread via the blood [Eklund et al., 1967]. In addition, several experiments performed in diverse laboratories have clearly shown that blood carries infectivity before the appearance of clinical signs of the disease [Brown et al., 2001]. In a recent study, Banks and colleagues reported that purified PrP^{Sc} can reach the brain exclusively by crossing the blood–brain barrier (BBB) without the involvement of peripheral nerves [Banks et al., 2004]. For these experiments, animals were perfused so that the solution containing labeled PrP^{Sc} was exposed exclusively to the brain through the BBB. The results indicated that a significant amount of the protein reached the brain intact, providing a possible route for prion neuro-invasion.

On the other hand, there are many results indicating that prion transport from the lymphoid system to the central nervous system (CNS) occurs along peripheral nerves in a manner that seems to depend on the expression of PrP^C [Blattler et al., 1997; Glatzel and Aguzzi, 2000]. The innervation pattern of lymphoid organs is mainly sympathetic, and sympathectomy delays the transport of prions from lymphatic organs to the thoracic spinal cord [Glatzel et al., 2001], which is the entry site of sympathetic nerves to the CNS. However, the mechanisms associated with the transport of the infectious agent in peripheral nerves are far from clear. For example, it is not known how prions are transferred from the lymphoid tissues to the nerves, especially considering that no connection between FDCs and nerve endings has been described. In addition, it is unclear how prions are actually transported within peripheral nerves. Axonal and nonaxonal transport mechanisms may be involved [Haik et al., 2004]. However, data showing that inhibitors of fast axonal transport did not alter prion infectivity in experimental animals argue against axonal transport as a putative mechanism of prion neuroinvasion [Kunzi et al., 2002]. A likely hypothesis is that the spread of PrP^{Sc} mainly involves direct exchange between plasma membranes of adjacent cells. Alternatively, it might be possible that the infectious agent spreads through the conversion of PrP^C associated with one cell terminal induced by PrP^{Sc} bound to the membrane of the neighboring cell. This model may explain why the rate of neural prion spread is so low and may not follow the canonical mechanisms of fast axonal transport.

Support for a role of sympathetic nerves in prion transport came from studies showing that infectivity titers in hyperinnervated spleens are at least two logs higher and show enhanced PrP^{Sc} accumulations compared with control mice [Clarke and Kimberlin, 1984]. However, experiments with denervated mice show that, although the onset of the disease was delayed, animals still got ill [Glatzel et al., 2001]. This finding suggests that an alternative route of entry exists, which may become uncovered by the absence of sympathetic fibers. Entry through the vagal nerve has been proposed in studies of the dynamics of vacuolation following oral and intraperitoneal challenge with prions [Baldauf et al., 1997; Beekes et al., 1998].

After getting into the brain, prions appear to spread relatively rapidly to different brain areas. This finding is more easily explained if prions get into the brain through the BBB because, in this case, multiple entry points would exist, which is probably not the case in the peripheral-nerve model.

6.4   Concluding remarks

Despite much effort to understand the involvement of the immune system and lymphoid organs on the transport of prions to the brain, the mechanism by which the infectious agent gets into the CNS is, for the most part, unknown. Little is known about the transit of prions through the gastrointestinal tract and the mechanism by which they permeate the intestinal barrier. Perhaps the most unambiguous aspect is the important role of the Peyer's patches in this process. Much more research has been done around the peripheral replication phase of prions in lymphoid organs, particularly the spleen. However, there is still a great deal of controversy regarding which cell types are necessary for prion transport and replication (Figure 6.2). The available evidence suggests that FDCs are the most likely peripheral reservoir for prions, but lymphocytes also play a major role in the transport of the infectious agent toward the brain. The last step in the transport from the lymphoid tissues to the brain most likely involves peripheral sympathetic nerves. However, recent evidence also points to the possibility that infectious proteins directly cross the BBB to reach the CNS. A better understanding of the diverse routes implicated and the tissues, organs, and cells participating in prion neuroinvasion could lead to novel targets for therapeutic intervention and new alternatives for early diagnosis.

References ***

**Aguzzi, A., Prions and the immune system: a journey through gut, spleen, and nerves, *Adv. Immunol.*, 81, 123–171, 2003.

**Aguzzi, A. et al., Immune system and peripheral nerves in propagation of prions to CNS, *Br. Med. Bull.*, 66, 141–159, 2003.

**Aucouturier, P. and Carnaud, C., The immune system and prion diseases: a relationship of complicity and blindness, *J. Leukoc. Biol.*, 72, 1075–1083, 2002.

**Aucouturier, P. et al., Prion diseases and the immune system, *Clin. Immunol.*, 96, 79–85, 2000.

Baldauf, E., Beekes, M., and Diringer, H., Evidence for an alternative direct route of access for the scrapie agent to the brain bypassing the spinal cord, *J. Gen. Virol.*, 78, 1187–1197, 1997.

Banchereau, J. and Steinman, R.M., Dendritic cells and the control of immunity, *Nature*, 392, 245–252, 1998.

*Banks, W.A. et al., Passage of murine scrapie prion protein across the mouse vascular blood–brain barrier, *Biochem. Biophys. Res. Commun.*, 318, 125–130, 2004. (An interesting study that reports for the first time direct evidence for prion uptake by the blood–brain barrier.)

Beekes, M., McBride, P.A., and Baldauf, E., Cerebral targeting indicates vagal spread of infection in hamsters fed with scrapie, *J. Gen. Virol.*, 79 (Pt 3), 601–607, 1998.

Blattler, T. et al., PrP-expressing tissue required for transfer of scrapie infectivity from spleen to brain, *Nature*, 389, 69–73, 1997.

Brandner, S., Klein, M.A., and Aguzzi, A., A crucial role for B cells in neuroinvasive scrapie, *Transfus. Clin. Biol.*, 6, 17–23, 1999.

---

* Highlight primary articles of outstanding importance and quality, including a short description of the findings.

** Highlights comprehensive review articles similar to the topic of this chapter.

*Brown, K.L. et al., Scrapie replication in lymphoid tissues depends on prion protein-expressing follicular dendritic cells, *Nat. Med.*, 5, 1308–1312, 1999. (A study showing the importance of follicular dendritic cells in prion peripheral replication.)

Brown, P., Cervenakova, L., and Diringer, H., Blood infectivity and the prospects for a diagnostic screening test in Creutzfeldt-Jakob disease, *J. Lab. Clin. Med.*, 137, 5–13, 2001.

Cashman, N.R. et al., Cellular isoform of the scrapie agent protein participates in lymphocyte activation, *Cell*, 61, 185–192, 1990.

Clarke, M.C. and Kimberlin, R.H., Pathogenesis of mouse scrapie: distribution of agent in the pulp and stroma of infected spleens, *Vet. Microbiol.*, 9, 215–225, 1984.

Dickinson, A.G. et al., Mitogenic stimulation of the host enhances susceptibility to scrapie, *Nature*, 272, 54–55, 1978.

Eklund, C.M., Kennedy, R.C., and Hadlow, W.J., Pathogenesis of scrapie virus infection in the mouse, *J. Infect. Dis.*, 117, 15–22, 1967.

*Fraser, H. and Dickinson, A.G., Pathogenesis of scrapie in the mouse: the role of the spleen, *Nature*, 226, 462–463, 1970. (One of the classical studies that indicated the important role of spleen in prion neuroinvasion.)

Frigg, R. et al., Scrapie pathogenesis in subclinically infected B-cell-deficient mice, *J. Virol.*, 73, 9584–9588, 1999.

Ghosh, S., Mechanism of intestinal entry of infectious prion protein in the pathogenesis of variant Creutzfeldt-Jakob disease, *Adv. Drug Deliv. Rev.*, 56, 915–920, 2004.

Glatzel, M. and Aguzzi, A., PrP(C) expression in the peripheral nervous system is a determinant of prion neuroinvasion, *J. Gen. Virol.*, 81, 2813–2821, 2000.

*Glatzel, M. et al., Sympathetic innervation of lymphoreticular organs is rate limiting for prion neuroinvasion, *Neuron*, 31, 25–34, 2001. (A study demonstrating the connection between peripheral nerves and lymphoid tissue in prion neuroinvasion.)

Haik, S., Faucheux, B.A., and Hauw, J.J., Brain targeting through the autonomous nervous system: lessons from prion diseases, *Trends Mol. Med.*, 10, 107–112, 2004.

Heppner, F.L. et al., Transepithelial prion transport by M cells, *Nat. Med.*, 7, 976–977, 2001.

Jeffrey, M. et al., Sites of prion protein accumulation in scrapie-infected mouse spleen revealed by immuno-electron microscopy, *J. Pathol.*, 191, 323–332, 2000.

Kimberlin, R.H. and Walker, C.A., Pathogenesis of mouse scrapie: dynamics of agent replication in spleen, spinal cord and brain after infection by different routes, *J. Comp. Pathol.*, 89, 551–562, 1979.

*Klein, M.A. et al., A crucial role for B cells in neuroinvasive scrapie, *Nature*, 390, 687–690, 1997. (A pioneer study using immunodeficient animals to understand the contribution of the immune system in prion propagation.)

Klein, M.A. et al., PrP expression in B lymphocytes is not required for prion neuroinvasion, *Nat. Med.*, 4, 1429–1433, 1998.

Kunzi, V. et al., Unhampered prion neuroinvasion despite impaired fast axonal transport in transgenic mice overexpressing four-repeat tau, *J. Neurosci.*, 22, 7471–7477, 2002.

Kuroda, Y. et al., Creutzfeldt-Jakob disease in mice: persistent viremia and preferential replication of virus in low-density lymphocytes, *Infect. Immunol.*, 41, 154–161, 1983.

Montrasio, F. et al., Impaired prion replication in spleens of mice lacking functional follicular dendritic cells, *Science*, 288, 1257–1259, 2000.

Neutra, M.R., Frey, A., and Kraehenbuhl, J.P., Epithelial M cells: gateways for mucosal infection and immunization, *Cell*, 86, 345–348, 1996.

Outram, G.W., Dickinson, A.G., and Fraser, H., Developmental maturation of susceptibility to scrapie in mice, *Nature*, 241, 536–537, 1973.

Outram, G.W., Dickinson, A.G., and Fraser, H., Reduced susceptibility to scrapie in mice after steroid administration, *Nature*, 249, 855–856, 1974.

Press, C.M., Heggebo, R., and Espenes, A., Involvement of gut-associated lymphoid tissue of ruminants in the spread of transmissible spongiform encephalopathies, *Adv. Drug Deliv. Rev.*, 56, 885–899, 2004.

Wadsworth, J.D. et al., Tissue distribution of protease resistant prion protein in variant Creutzfeldt-Jakob disease using a highly sensitive immunoblotting assay, *Lancet*, 358, 171–180, 2001.

Wells, G.A. et al., Infectivity in the ileum of cattle challenged orally with bovine spongiform encephalopathy, *Vet. Rec.*, 135, 40–41, 1994.

chapter seven

Neurodegeneration in prion diseases

Although transmissible spongiform encephalopathies (TSEs) are diseases without a unique etiology or clinical profile, the characteristics of brain damage are similar. Extensive spongiform degeneration, a large extent of neuronal loss, synaptic alterations, atypical brain inflammation, and accumulation of protein aggregates are typical features of TSE [Castilla et al. 2004]. This chapter describes the features of brain degeneration, the putative relationship of the prion protein and its structural conversion with brain damage, and the signaling pathways involved in TSE neurodegeneration.

7.1   Characteristics of brain degeneration

One intriguing feature of prion diseases is that histopathological changes are restricted to the central nervous system (CNS) despite the trafficking of pathological prion proteins (PrP^{Sc}) along the peripheral nervous system and the detection of PrP^{Sc} in lymphoid organs (see Chapter 6). The typical neuropathological alterations in TSEs are vacuolation of the gray matter, prominent neuronal loss, atypical brain inflammation, synaptic dysfunction, and a variable degree of cerebral accumulation of prion protein (PrP) aggregates [Budka et al., 1995; MacDonald et al., 1996; Wells, 1993]. Besides deposition of PrP^{Sc}, all others histopathological features of prion diseases can be found in other acute and chronic CNS injuries.

The most specific of the brain abnormalities is certainly the vacuolation, giving to the brain the appearance of a sponge and hence the name of spongiform encephalopathy. The spongiform degeneration consists of diffuse or focally clustered, small, round vacuoles that may become confluent [Wells, 1993]. Despite the widely recognized presence of brain vacuolation, the role of spongiform degeneration in TSE's clinical symptoms and in disease onset has been little studied and is, for the most part, unclear. Brain inflammation is prominent in TSEs, but it is rather unusual, because it consists mainly of activation of astrocytes and microglia, with very little or no lymphocyte infiltration [Betmouni et al., 1996]. Again, the contribution of inflammation to brain dysfunction and disease onset remains unclear. At early stages of scrapie infection in mice, synaptic abnormalities associated with deposition of abnormal PrP have been described in mice long before neuronal death [Jeffrey et al., 2000]. Recently, synaptic dysfunction has been linked to behavioral changes, but the mechanisms underlying it remain to be established [Cunningham et al., 2003]. A role for PrP in synaptic function has also been suggested in studies performed in PrP knockout mice [Collinge et al., 1994; Lledo et al., 1996]. Normal cellular protein (PrP^C) has been shown to interact with other presynaptic proteins, including synaptophysin, and hence an interference of this interaction due to conversion of PrP^C into PrP^{Sc} may play a role in the disease [Keshet et al., 2000]. Our recent findings indicate that, at early stages of the disease, PrP^{Sc} accumulates in lipid rafts, producing a detachment of caveolin and synaptophysin from these membrane domains [Russelakis-Carneiro et al., 2004]. The identification of early events in the pathogenesis of prion diseases is crucial for the development of therapeutic strategies aimed to avoid brain damage and disease.

Neuronal loss is a salient feature of prion diseases [Gray et al., 1999b]. Although recent evidence coming from longitudinal studies with animal models has suggested that the clinical signs appear long before any detectable neuronal loss, most of the studies on brain degeneration in TSE have been focused on understanding the mechanism of neuronal death and the role of PrP^{Sc}. Several studies, in humans and in mice models, of diverse types of prion diseases (infectious models, hereditary models with mutated PrPs, and transgenic models overexpressing wild-type PrP) indicate that neuronal dysfunction and death occur through a specific mechanism denominated as apoptosis [Dorandeu et al., 1998; Ferrer, 1999; Giese et al., 1995; Gray et al., 1999a; Jesionek-Kupnicka et al., 1999; Lucassen et al., 1995; Peretz et al., 1997]. Transgenic mice expressing a PrP insertional mutation showed massive apoptosis of granule cells in the cerebellum, including nuclear condensation and fragmentation, internucleosomal cleavage of DNA, and caspase-3 activation [Chiesa et al., 2000]. Similarly, transgenic mice overexpressing wild-type [Westaway et al., 1994] or mutant PrP [Hsiao et al., 1991] also showed similar mechanisms of neuronal death and induced spontaneous neurological disease with spongiform degeneration [Prusiner, 1996a].

The time sequence of the different brain alterations and their association with clinical disease is not entirely known and most likely is different for distinct prion strains. However, longitudinal studies in animal models indicate a possible sequence of events (Figure 7.1), beginning with PrP^{Sc} formation and ending with the clinical disease.

7.2   Is PrP^{Sc} the cause of TSE neurodegeneration?

Compelling evidence links PrP^{Sc} with the pathogenesis of prion diseases [Prusiner, 1998]. First, all individuals who develop TSE contain PrP^{Sc} in their brains and, indeed, the presence of the misfolded protein is considered

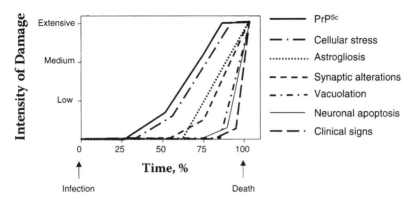

Figure 7.1 Time-course progression of prion disease pathogenesis. After inoculation of mice with murine scrapie, it is possible to detect PrPSc deposition several weeks postinfection but before the appearance of the clinical symptoms. Soon after and closely following PrPSc formation, it is possible to detect signals of cellular stress, in particular the up-regulation of ER chaperones. Subsequently, local brain inflammation can be detected in the brain areas showing PrPSc deposition, reflected in astrogliosis and the production of proinflammatory compounds. At this stage, it is also possible to detect deficiencies in synaptic markers. At late stages of the disease, brain vacuolation and cell death by apoptosis occur, leading to irreversible damage of the brain. The first clinical signs of the disease appear at this time. The graph is based on data reported previously (Williams, A. et al., 1997; Hetz, C. et al., 2003; Russelakis-Carneiro, M. et al., 2004).

essential for definitive diagnosis [Prusiner, 1998]. Second, in general there is a very good temporal and anatomical correlation between PrPSc accumulation and the appearance of neuropathology and clinical symptoms [DeArmond et al., 1987]. In addition, all mutations associated with familial forms of prion diseases are located in the PrP gene and, in general, these favor the formation of the misfolded protein [Prusiner and Scott, 1997]. Moreover, transgenic animal models have been generated in which overexpression of mutant PrP was shown to induce spontaneous neurological disease with spongiform degeneration [Prusiner, 1996b]. These abnormalities could be transmitted to normal mice by inoculation of the disease-affected brain homogenate. Finally, nanomolar concentrations of highly purified PrPSc are highly neurotoxic *in vitro*, inducing cellular apoptosis, and this activity is very specific for the β-sheet misfolded form [Hetz et al., 2003].

However, a number of results argue against a simple relationship between PrPSc and tissue degeneration [Chiesa and Harris, 2001]. First, despite the generally good correlation between PrPSc and neurodegeneration, it is possible to see in some brain areas a strong PrPSc staining without any evident neuropathological damage [Parchi et al., 1995]. Although it is possible that other neuroprotective factors might be protecting certain specific types of neurons, these findings suggest that PrPSc might not be intrinsically neurotoxic. Perhaps the most persuasive arguments against a simple relationship between PrP^{Sc} and neurodegeneration come from studies using transgenic animal models, in which there are examples of both neuropathology in the absence of detectable PrP^{Sc} and lack of brain damage in animals bearing high levels of PrP^{Sc}. (For a review, see [Chiesa and Harris, 2001].)

An interesting possibility that has been studied recently is that PrP^{Sc}'s neurotoxic effect is mediated through PrP^{C}. Mallucci and coworkers have found that depleting endogenous neuronal PrP^{C} in mice with established neuroinvasive prion infection reversed early spongiform change and prevented neuronal loss and progression to clinical disease [Mallucci et al., 2003]. This occurred despite the accumulation of extraneuronal PrP^{Sc} to levels seen in terminally ill wild-type animals. Thus, the propagation of nonneuronal PrP^{Sc} seems not to be pathogenic. In a related study, Brandner and colleagues showed that PrP-null brain tissue surrounding prion-infected *Prnp*$^{+/+}$ neurografts does not develop prion neuropathological changes [Brandner et al., 1996]. In both cases, PrP^{Sc} seems unable to trigger neuronal death in the nearby PrP knockout cells. These results suggest that PrP^{Sc} would need the presence of PrP^{C} to be toxic. This interpretation finds support in recent data from Solforosi et al. suggesting that neurodegeneration could be directly triggered through a cross-linking of PrP^{C} by a monoclonal antibody [Solforosi et al., 2004]. The extrapolation of these findings is that oligomeric PrP^{Sc} could be the activator of a PrP^{C}-mediated signaling pathway, but not the direct causal molecule of neuronal loss, since other molecules with similar PrP^{C} cross-linking activity as monoclonal antibodies could produce the same effect.

Another unclear aspect regarding PrP neurotoxicity is the nature of the toxic molecule. There is evidence indicating that diverse species could be the neurotoxic form, including mature PrP^{Sc}, conformational intermediates between PrP^{C} and PrP^{Sc}, cytosolic PrP^{Sc}-like forms (see below), protease-sensitive PrP^{Sc} isoforms, and transmembrane forms of PrP. (For a review, see [Chiesa and Harris, 2001].) Moreover, it is likely that in different TSEs, distinct PrP species (or combinations of species) might be causing the pathology.

7.3   Mechanism of neuronal apoptosis

Apoptosis is a programmed form of cell death that plays a central role during development and homeostasis of multicellular organisms and has been implicated in a number of pathological conditions [Jacobson et al., 1997; Reed, 2002; Vaux and Korsmeyer, 1999]. The central executioner molecules of apoptosis are a large family of cysteine proteases known as caspases [Hengartner, 2000; Takahashi, 1999]. Based on structural similarities, substrate preference, and role in the apoptotic pathway, caspases have been divided into initiators (such as caspase-8 and caspase-9), downstream executor caspases (such as caspase-3), and inflammatory caspases (such as caspase-1). Activation of caspase-dependent apoptosis may be initiated by activation of death receptors or by mitochondrial stress [Budihardjo et al., 1999] (Figure 7.2). Recently, another apoptotic-regulatory pathway has been described, in which the induction of endoplasmic reticulum (ER) stress due to alteration of calcium homeostasis or accumulation of misfolded proteins triggers the activation of an ER-resident caspase, termed caspase-12 [Nakagawa et al., 2000] (Figure 7.2). In the last 3 years, several reports have linked the ER-stress apoptosis pathway to diverse neurodegenerative diseases related to protein misfolding and aggregation.

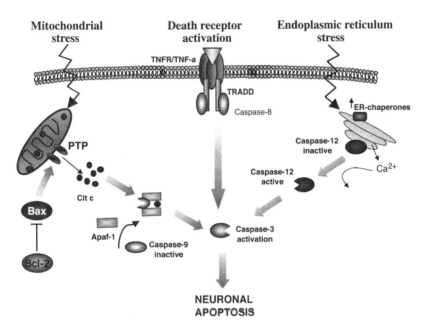

Figure 7.2 Signaling pathways involved in neuronal apoptosis. At least three different pathways have been reported to lead to cellular apoptosis. The so-called extrinsic pathway begins with the activation of cell-surface death receptors by binding to a ligand. This interaction leads to oligomerization of the receptor, which then interacts with adaptor proteins to recruit and activate the initiator caspase-8. The mitochondrial pathway, triggered by insults that induce stress in this organelle or DNA damage, consists of the opening of the mitochondrial permeability transition pore (PTP), leading to the release of cytochrome c to the cytosol. Cytochrome c forms a complex with the adaptor protein apaf-1 and pro-caspase-9, leading to the proteolytic activation of this enzyme. Finally, there is the ER pathway, in which ER damage or accumulation of misfolded proteins induces the release of calcium from the ER. The initial response to ER stress consists of the up-regulation of ER chaperones (i.e., Grp58, Grp78/Bip, Grp94, heat-shock proteins), but when the damage persists, the ER-resident caspase-12 becomes activated. The active forms of caspases 8, 9, or 12 activate the executor caspase-3, leading to apoptosis.

7.4   Neuronal apoptosis in TSEs involves the ER-stress pathway

Purified PrP^{Sc} from scrapie-infected brains was shown to induce ER stress in neuroblastoma cells in culture. This is reflected by the sustained release of ER calcium and the induction of the unfolded protein response (UPR) associated with the up-regulation of several ER chaperones of the glucose-regulated chaperone protein family (termed Grps) [Hetz et al., 2003]. It is known that the first response to ER stress consists of an antiapoptotic pathway related to the up-regulation of different chaperones and folding enzymes in an attempt to correct or remove misfolded proteins, thereby minimizing their accumulation and toxicity. When the apoptotic stimulus is persistent, a proapoptotic response is triggered, leading to cell death, which is mediated by the activation of the ER-resident caspase-12 (or caspase-4 in humans). After treatment of cells with PrP^{Sc}, caspase-12 was specifically activated as a consequence of ER stress, which in turn activated caspase-3, the central apoptosis executer [Hetz et al., 2003]. These findings in cell culture are supported by histological and biochemical analysis of brains from scrapie-sick mice and from humans affected by sporadic Creutzfeldt-Jakob disease (sCJD) and variant CJD (vCJD). Activation of caspase-12 and induction of ER-stress-inducible chaperones was reported in brain areas exhibiting extensive neuronal death [Hetz et al., 2003; Yoo et al., 2002]. Furthermore, longitudinal studies in experimental models of scrapie in rodents showed that up-regulation of the ER chaperone protein Grp58 was closely associated with PrP^{Sc} accumulation and that the expression of this chaperone modulates the neurotoxic activity of misfolded PrP [Hetz et al., 2005].

There is much other indirect evidence linking prion pathogenesis to ER alterations, namely:

- The stimulation of retrograde transport toward ER increases the accumulation of PrP^{Sc} in prion-infected neuroblastoma cells [Beranger et al., 2002].
- Several PrP mutants associated with inherited prion diseases are retained in this organelle [Jin et al., 2000; Singh et al., 1997].
- Expression of mutant PrP induces aggregation of wild-type PrP in the ER [Gu et al., 2003].
- Scrapie-infected neuroblastoma cells are more susceptible to ER stress [Hetz et al., 2003].

It remains to be determined whether, in these experimental systems, ER stress is directly associated with each pathological condition.

Interestingly, several ER chaperones, including Bip/Grp78, calnexin, and protein disulfide isomerase, have been shown to interact with mutant PrP [Capellari et al., 2000]. However, no PrP^{Sc} accumulation in the ER has been detected in sporadic or infectious forms of TSE, making it difficult to relate mechanistically the generation of PrP^{Sc} to the induction of ER stress. Only one report suggested that, in scrapie-infected cells, PrP could be detected in the cytosol, associated with Golgi markers [Taraboulos et al., 1990], but this observation has not been corroborated. It is likely that the relationship between PrP^{Sc} and ER stress is mediated by changes in calcium homeostasis. Indeed, our *in vitro* experiments suggest that the first abnormality observed after exposure of the cells to purified PrP^{Sc} is the release of calcium from the ER, which occurs within minutes after treatment with the pathological protein [Hetz et al., 2003]. Also, it has been reported that chronically infected cells present modifications in calcium homeostasis [Kristensson et al., 1993; Wong et al., 1996]. Severe ER-calcium alterations are also reflected in the accumulation of misfolded proteins in the ER and the induction of the UPR response, because calcium is essential for the proper folding of proteins in the ER. (See review in [Brostrom and Brostrom, 2003].) Thus, the relationship between prion misfolding and ER-stress induction could be an indirect effect via calcium signaling. If calcium signaling is essential for PrP^{Sc} toxicity, it would be important to elucidate the mechanism by which PrP^{Sc} induces the release of ER calcium.

7.5   A role for the proteasome in TSE pathogenesis?

Protein folding is a highly regulated process mediated by diverse chaperones and foldases. In all protein-misfolding disorders (see Chapter 11), clearance of the misfolded proteins plays a key role in controlling its accumulation and pathogenicity. It has been estimated that approximately 10% of nascent PrP molecules are spontaneously misfolded and targeted to the ER-associated degradation pathway (termed ERAD) [Ma and Lindquist, 2002; Yedidia et al., 2001]. During this process, PrP is deglycosylated and ubiquitinated for proteasomal degradation. ER chaperones, like Grp78/Bip, are implicated in the process of recognition for proteasomal degradation, playing a role in the maintenance of PrP quality control. A failure of the proteasome system during aging has been proposed as a possible mechanism for the onset of sporadic CJD.

It has been described that, after proteasome inhibition with different compounds, fully mature PrP is accumulated in the cytoplasm and exhibits some of the biochemical properties of PrP^{Sc}, such as insolubility in nonionic detergents and resistance to protease degradation [Ma and Lindquist, 2002; Yedidia et al., 2001]. Interestingly, the accumulation of PrP in the cytoplasm, following proteasome inhibition in PrP-overexpressing cells, was shown to cause cellular death by apoptosis. After removal of the proteasome inhibitors, the accumulation of PrP in the cytosol continued, suggesting that a self-perpetuating (infectious-like) mechanism could be involved [Ma and Lindquist, 1999]. However, it has been suggested that these results could be produced by an artificial increase in PrP mRNA levels because of an unspecific effect of those inhibitors in the viral promoter used to overexpress PrP [Drisaldi et al., 2003]. In addition, another publication suggested that cytosolic PrP is not toxic in primary neuronal cultures and does not show any of the

PrP^{Sc}-like properties that are observed in neuroblastoma cell lines [Roucou et al., 2003]. Moreover, the later results reinforced the notion that cellular PrP^C has an antiapoptotic activity when it is expressed in the cytosol [Roucou et al., 2003]. (See Chapter 4 for a description of putative biological functions of PrP^C.) A more detailed analysis of proteasome dysfunction must be performed in more-relevant disease-associated models or in individuals affected with naturally occurring forms of prion diseases to solve this controversy.

Transgenic mice expressing a cytosolic PrP mimicking the PrP form generated through the ERAD process developed severe ataxia with cerebellar degeneration and gliosis. It was postulated that, in conditions in which the proteasome ability to degrade PrP is compromised, like cellular stress and events associated with aging, accumulation of cytosolic PrP may promote neuronal degeneration. A similar intracellular accumulation in the ER has been observed in cell lines overexpressing mutant PrP (E200K y D178N), which is related to familial CJD in humans [Negro et al., 2001]. Another mutant PrP associated with GSS, the PrP Q217R, was also shown to be accumulated in the ER [Singh et al., 1997] and degraded by the proteasome system [Jin et al., 2000]. It is possible to speculate that the mechanism by which proteasomal dysfunction induces cellular death when PrP is accumulated in the cytoplasm could be associated with the accumulation of misfolded and immature proteins in the ER (i.e., PrP and other normal cellular proteins subjected to ERAD) [Dimcheff et al., 2003]. This protein "traffic jam" would activate the cell-death pathways involved in the ER-stress response. In this case, cytosolic PrP may not be toxic *per se*, but it might promote the overload of proteasome with highly protease-resistant proteins, delaying the normal process of protein degradation through ERAD.

7.6   Concluding remarks

Cerebral damage observed in TSEs is characterized by spongiform degeneration of the brain, synaptic alterations, brain inflammation, neuronal death, and the accumulation of aggregated PrP^{Sc}. It is unclear which of these alterations triggers the disease onset, but it is likely that several of them are involved in the progressive tissue deterioration. Also unclear is the involvement of PrP structural conversion on brain degeneration, but *in vitro* studies have shown that ER stress induced by treatment with purified PrP^{Sc} causes neuronal death. ER stress leads to activation of caspase-12, which in turn activates the apoptosis executer, caspase-3. The cellular protective signaling pathways normally involved in the elimination of misfolded protein during synthesis and maturation of proteins play a key role in controlling the cellular fate. This response consists of the specific up-regulation of ER chaperones in an attempt to correct or remove the misfolded protein. However, this mechanism of defense may turn out to have a negative consequence by overloading the proteasomal degradation system, resulting in dysfunction of this important cellular process and contributing to cell damage.

References* **

Beranger, F. et al., Stimulation of PrP(C) retrograde transport toward the endoplasmic reticulum increases accumulation of PrP(Sc) in prion-infected cells, *J. Biol. Chem.*, 277, 38972–38977, 2002.

Betmouni, S., Perry, V.H., and Gordon, J.L., Evidence for an early inflammatory response in the central nervous system of mice with scrapie, *Neuroscience*, 74, 1–5, 1996.

Brandner, S. et al., Normal host prion protein necessary for scrapie-induced neurotoxicity, *Nature*, 379, 339–343, 1996.

Brostrom, M.A. and Brostrom, C.O., Calcium dynamics and endoplasmic reticular function in the regulation of protein synthesis: implications for cell growth and adaptability, *Cell Calcium*, 34, 345–363, 2003.

Budihardjo, I. et al., Biochemical pathways of caspase activation during apoptosis, *Ann. Rev. Cell Dev. Biol.*, 15, 269–290, 1999.

Budka, H. et al., Neuropathological diagnostic criteria for Creutzfeldt-Jakob disease (CJD) and other human spongiform encephalopathies (prion diseases), *Brain Pathol.*, 5, 459–466, 1995.

Capellari, S. et al., Effect of the E200K mutation on prion protein metabolism: comparative study of a cell model and human brain, *Am. J. Pathol.*, 157, 613–622, 2000.

**Castilla, J., Hetz, C., and Soto, C., Molecular mechanism of neurotoxicity of pathological prion protein, *Curr. Mol. Med.*, 4, 397–403, 2004.

Chiesa, R. et al., Accumulation of protease-resistant prion protein (PrP) and apoptosis of cerebellar granule cells in transgenic mice expressing a PrP insertional mutation, *Proc. Natl. Acad. Sci. USA*, 97, 5574–5579, 2000.

**Chiesa, R. and Harris, D.A., Prion diseases: what is the neurotoxic molecule? *Neurobiol. Dis.*, 8, 743–763, 2001.

Collinge, J. et al., Prion protein is necessary for normal synaptic function, *Nature*, 370, 295–297, 1994.

Cunningham, C. et al., Synaptic changes characterize early behavioural signs in the ME7 model of murine prion disease, *Eur. J. Neurosci.*, 17, 2147–2155, 2003.

DeArmond, S.J. et al., Changes in the localization of brain prion proteins during scrapie infection, *Neurology*, 37, 1271–1280, 1987.

Dimcheff, D.E., Portis, J.L., and Caughey, B., Prion proteins meet protein quality control, *Trends Cell Biol.*, 13, 337–340, 2003.

Dorandeu, A. et al., Neuronal apoptosis in fatal familial insomnia, *Brain Pathol.*, 8, 531–537, 1998.

Drisaldi, B. et al., Mutant PrP is delayed in its exit from the endoplasmic reticulum, but neither wild-type nor mutant PrP undergoes retrotranslocation prior to proteasomal degradation, *J. Biol. Chem.*, 278, 21732–21743, 2003.

Ferrer, I., Nuclear DNA fragmentation in Creutzfeldt-Jakob disease: does a mere positive *in situ* nuclear end-labeling indicate apoptosis? *Acta Neuropathol. (Berl)*, 97, 5–12, 1999.

Giese, A. et al., Neuronal cell death in scrapie-infected mice is due to apoptosis, *Brain Pathol.*, 5, 213–221, 1995.

* Highlights primary articles of outstanding importance and quality, including a short description of the findings.
** Highlights comprehensive review articles similar to the topic of this chapter.

Gray, F. et al., Neuronal apoptosis in human prion diseases, *Bull. Acad. Natl. Med.*, 183, 305–320, 1999a.

Gray, F. et al., Neuronal apoptosis in Creutzfeldt-Jakob disease, *J. Neuropathol. Exp. Neurol.*, 58, 321–328, 1999b.

Gu, Y. et al., Mutant prion protein-mediated aggregation of normal prion protein in the endoplasmic reticulum: Implications for prion propagation, *J. Neurochem.*, 84, 10–22, 2003.

Hengartner, M.O., The biochemistry of apoptosis, *Nature*, 407, 770–776, 2000.

*Hetz, C. et al., Caspase-12 and endoplasmic reticulum stress mediate neurotoxicity of pathological prion protein, *EMBO J.*, 22, 5435–5445, 2003. (A study showing *in vitro* and *in vivo* data suggesting that PrP^{Sc} may induce neuronal apoptosis by the ER-stress-mediated pathway.)

Hetz, C. et al., The disulfide isomerase Grp58 is a neuroprotective factor against prion replication, *J. Neurosci.*, 25, 2793–2802, 2005.

Hsiao, K. et al., Spontaneous neurodegeneration in transgenic mice with prion protein codon 101 proline–leucine substitution, *Ann. N.Y. Acad. Sci.*, 640, 166–170, 1991.

Jacobson, M.D., Weil, M., and Raff, M.C., Programmed cell death in animal development, *Cell*, 88, 347–354, 1997.

Jeffrey, M. et al., Synapse loss associated with abnormal PrP precedes neuronal degeneration in the scrapie-infected murine hippocampus, *Neuropathol. Appl. Neurobiol.*, 26, 41–54, 2000.

Jesionek-Kupnicka, D. et al., Neuronal loss and apoptosis in experimental Creutzfeldt-Jakob disease in mice, *Folia Neuropathol.*, 37, 283–286, 1999.

Jin, T. et al., The chaperone protein BiP binds to a mutant prion protein and mediates its degradation by the proteasome, *J. Biol. Chem.*, 275, 38699–38704, 2000.

Keshet, G.I. et al., The cellular prion protein colocalizes with the dystroglycan complex in the brain, *J. Neurochem.*, 75, 1889–1897, 2000.

Kristensson, K. et al., Scrapie prions alter receptor-mediated calcium responses in cultured cells, *Neurology*, 43, 2335–2341, 1993.

Lledo, P.M. et al., Mice deficient for prion protein exhibit normal neuronal excitability and synaptic transmission in the hippocampus, *Proc. Natl. Acad. Sci. USA*, 93, 2403–2407, 1996.

Lucassen, P.J. et al., Detection of apoptosis in murine scrapie, *Neurosci. Lett.*, 198, 185–188, 1995.

*Ma, J. and Lindquist, S., De novo generation of a PrPSc-like conformation in living cells, *Nat. Cell Biol.*, 1, 358–361, 1999. (This, as well as the following study from the same group, describes the generation of a PrP^{Sc}-like form in the cytosol by culturing cells under reducing and deglycosylation conditions.)

*Ma, J. and Lindquist, S., Conversion of PrP to a self-perpetuating PrPSc-like conformation in the cytosol, *Science*, 298, 1785–1788, 2002.

MacDonald, S.T., Sutherland, K., and Ironside, J.W., Prion protein genotype and pathological phenotype studies in sporadic Creutzfeldt-Jakob disease, *Neuropathol. Appl. Neurobiol.*, 22, 285–292, 1996.

*Mallucci, G. et al., Depleting neuronal PrP in prion infection prevents disease and reverses spongiosis, *Science*, 302, 871–874, 2003. (A very important study showing that, in conditional knockout animals, removal of PrP expression resulted in reversion of disease pathology.)

Nakagawa, T. et al., Caspase-12 mediates endoplasmic reticulum specific apoptosis and cytotoxicity by amyloid-beta, *Nature*, 403, 98–103, 2000.

Negro, A. et al., The metabolism and imaging in live cells of the bovine prion protein in its native form or carrying single amino acid substitutions, *Mol. Cell Neurosci.*, 17, 521–538, 2001.

Parchi, P. et al., Regional distribution of protease-resistant prion protein in fatal familial insomnia, *Ann. Neurol.*, 38, 21–29, 1995.

Peretz, D. et al., A conformational transition at the N terminus of the prion protein features in formation of the scrapie isoform, *J. Mol. Biol.*, 273, 614–622, 1997.

Prusiner, S.B., Human prion diseases and neurodegeneration, *Curr. Top. Microbiol. Immunol.*, 207, 1–17, 1996a.

Prusiner, S.B., Transgenetics of prion diseases, *Curr. Top. Microbiol. Immunol.*, 206, 275–304, 1996b.

Prusiner, S.B., Prions, *Proc. Natl. Acad. Sci. USA*, 95, 13363–13383, 1998.

Prusiner, S.B. and Scott, M.R., Genetics of prions, *Annu. Rev. Genet.*, 31, 139–175, 1997.

Reed, J.C., Apoptosis-based therapies, *Nat. Rev. Drug Discov.*, 1, 111–121, 2002.

*Roucou, X. et al., Cytosolic prion protein is not toxic and protects against Bax-mediated cell death in human primary neurons, *J. Biol. Chem.*, 278, 40877–40881, 2003. (An intriguing study showing a possible implication of PrP^C in neuroprotection by inhibiting the Bax-mediated apoptosis pathway.)

Russelakis-Carneiro, M., Hetz, C., Maundrell, K., and Soto, C., Accumulation of PrP in neuronal lipid rafts leads to a change in the subcellular localization of caveolin and synaptophysin: a putative mechanism of neurodegeneration in prion disease, *Am. J. Pathol.*, 165, 1839–1848, 2004.

Singh, N. et al., Prion protein aggregation reverted by low temperature in transfected cells carrying a prion protein gene mutation, *J. Biol. Chem.*, 272, 28461–28470, 1997.

*Solforosi, L. et al., Cross-linking cellular prion protein triggers neuronal apoptosis in vivo, *Science*, 303, 1514–1516, 2004. (A study providing evidence for a PrP signaling role in neuronal apoptosis.)

Takahashi, A., Caspase: executioner and undertaker of apoptosis, *Int. J. Hematol.*, 70, 226–232, 1999.

Taraboulos, A., Serban, D., and Prusiner, S.B., Scrapie prion proteins accumulate in the cytoplasm of persistently infected cultured cells, *J. Cell Biol.*, 110, 2117–2132, 1990.

Vaux, D.L. and Korsmeyer, S.J., Cell death in development, *Cell*, 96, 245–254, 1999.

Wells, G.A., Pathology of nonhuman spongiform encephalopathies: variations and their implications for pathogenesis, *Dev. Biol. Stand.*, 80, 61–69, 1993.

Westaway, D. et al., Degeneration of skeletal muscle, peripheral nerves, and the central nervous system in transgenic mice overexpressing wild-type prion proteins, *Cell*, 76, 117–129, 1994.

Williams, A. et al., PrP deposition, microglial activation, and neuronal apoptosis in murine scrapie, *Exp. Neurol.*, 144, 433–438, 1997.

Wong, K. et al., Decreased receptor-mediated calcium response in prion-infected cells correlates with decreased membrane fluidity and IP3 release, *Neurology*, 47, 741–750, 1996.

Yedidia, Y. et al., Proteasomes and ubiquitin are involved in the turnover of the wild-type prion protein, *EMBO J.*, 20, 5383–5391, 2001.

Yoo, B.C. et al., Overexpressed protein disulfide isomerase in brains of patients with sporadic Creutzfeldt-Jakob disease, *Neurosci. Lett.*, 334, 196–200, 2002.

chapter eight

The diagnosis problem and current tests

The unprecedented nature of the prion infectious agent and its unique mode of transmission present substantial complications for early diagnosis. So far, diagnosis of transmissible spongiform encephalopathies (TSEs) is only done by clinical examination, and in the case of cattle, diagnosis is done by biochemical tests that detect the prion agent in the brain postmortem [Soto, 2004]. No presymptomatic diagnosis is yet available for any of the human or animal TSEs. Although TSEs are neurodegenerative diseases, the infectious agent usually replicates and accumulates in peripheral tissues before it gets into the central nervous system (see Chapter 6). It is clear that prions can be associated with a number of peripheral tissues and biological fluids, and thus it should be possible to detect the agent in the periphery. This chapter describes in detail the current state of the art of prion diagnosis and the novel strategies being employed in an effort to develop a sensitive, early, and noninvasive diagnosis of these diseases.

8.1   Importance of early diagnosis

Although TSEs are rare diseases in humans, the proven transmission among humans and from cattle to humans — and the long silent period between infection to clinical disease — makes early diagnosis a high priority to minimize the further spreading of the disease [Soto, 2004]. The development of a highly sensitive, noninvasive, and early detection test for prions would certainly have an impact in many different fields, including blood banks, the food industry, human diagnosis, etc. (Figure 8.1).

Over the past few years, it has become clear that bovine spongiform encephalopathy (BSE) is a serious problem, and despite the imposition of stringent precautionary measures, the disease continues to spread in some European countries and has been recently reported in the U.S., Canada, and Japan, where it has caused serious economic damage to the beef industry [Donnelly et al., 2003; Nolen, 2004; Schiermeier, 2001]. Although several tests

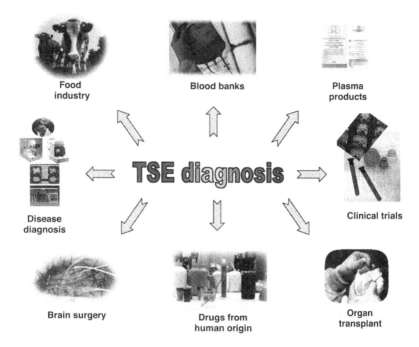

Food industry   Blood banks   Plasma products

TSE diagnosis

Disease diagnosis   Clinical trials

Brain surgery   Drugs from human origin   Organ transplant

Figure 8.1 Areas that may be impacted by a biochemical prion-detection assay. The currently available biochemical tests have been applied only in the beef industry to reduce the risk of infected animals reaching the food chain. It is likely that increasingly sensitive tests will expand the applications to the other areas described in this figure, including blood banks, plasma products, drugs produced from human tissues, organ transplant, brain surgery, etc.

have been developed to diagnose BSE in postmortem brain tissue, their usefulness in detecting presymptomatic animals is unclear [Schiermeier, 2001]. Thus, it is possible that infected animals not displaying clinical signs of BSE and testing negative for the available biochemical tests are still entering the food chain, imposing a serious risk to human health.

Presymptomatic detection of Creutzfeldt-Jakob disease (CJD) in living patients is not currently possible [Brown et al., 2001; Ingrosso et al., 2002; Schiermeier, 2001]. In a potential scenario where a substantial number of people may be incubating variant CJD (vCJD), the lack of a presymptomatic test has raised concerns about the iatrogenic propagation of the disease [Collinge, 1999; Wadsworth et al., 2001]. Iatrogenic spread of sporadic CJD (sCJD) has already taken the lives of several hundred people by transmission of prions via medical or surgical interventions using accidentally contaminated materials [Brown et al., 2000]. Iatrogenic transmission of sCJD has occurred in cases involving corneal transplants, implantation of electrodes in the brain, dura mater grafts, contaminated surgical instruments, and treatment with human growth hormone derived from cadaveric pituitaries (see Chapter 1) [Brown et al., 2000]. Although this proves that sCJD can be horizontally transmitted, the low incidence of iCJD indicates that the disease is not contagious in the traditional sense. However, in the case of vCJD, the concern is greater because the amount of pathological prion proteins (PrP^{Sc}) in peripheral lymphoreticular tissues is much higher than in sCJD [Wadsworth et al., 2001], indicating a different pattern of peripheral pathogenesis. Indeed, while no evidence of sCJD transmission by blood transfusion exists, two recent studies reported the first possible cases of vCJD acquired by blood transfusion [Llewelyn et al., 2004; Peden et al., 2004]. If the possibility of transmission of vCJD by blood transfusion is supported by more cases like these, it could lead to dramatic conse-quences for a vCJD epidemic, because it would indicate that blood carries infectivity several years before the onset of clinical symptoms. Because of the possibility of such a scenario, the development of a sensitive and presymptomatic blood test for CJD becomes a top priority.

Presymptomatic diagnosis is also very important for efficient treatment, as it would enable therapeutic intervention to be started at an early stage, before the appearance of clinical signs and the occurrence of permanent brain changes. It is known that at the time people exhibit clinical symptoms, the extent of brain damage is tremendous and most likely irreversible. The importance of early treatment is evident from the data available in animal models, which have shown that several compounds have effectively delayed the onset of the disease, but that none, so far, have shown a significant effect when treatment has been started at the symptomatic stage [Aguzzi et al., 2001; Dormont, 2003; Rossi et al., 2003].

8.2 *Difficulties of diagnosis*

Infectious diseases are, in general, relatively easy to diagnose biochemically, either by the immune reaction elicited in the body in response to the presence of the infectious agent or by the amplification of a nucleic acid specific to the agent using polymerase chain reaction (PCR). However, the protein nature of the infectious agent precludes the use of PCR, and no immune reaction has been detected in patients infected by prions.

The animal bioassays of infectivity remain the only methods available for directly measuring the infectious agent and are, by far, the most sensitive assays available for the detection of prions [Field and Shenton, 1972; Ingrosso et al., 2002]. Animal bioassays have been used extensively in TSE research and diagnostic testing. However, bioassays are severely limited in their widespread use by the length of time it takes to obtain results (several months to years) and the species-barrier effect.

PrP^{Sc} is not only the major component of the infectious agent and the most likely cause of TSE; it is also the only validated surrogate marker for the disease [Ingrosso et al., 2002; Prusiner, 1998]. Therefore, identification of PrP^{Sc} in human or animal tissues is considered key for TSE diagnosis. The main problem with a diagnosis based on detecting PrP^{Sc} is that the patho-logical form of PrP is abundant only at late stages of the disease in the primary target organ: the brain. However, infectivity studies have shown that prions are also present in lower amounts in peripheral tissues, such as lymphoid organs and blood, even at early stages during the presymptomatic period [Aguzzi, 2000; Brown, 2000; Brown et al., 2001]. The aim should be not only to detect prions in the brain in early presymptomatic cases, but also to generate a test for diagnosing living animals and humans. For this purpose, a tissue other than brain is required, and to create an easier, noninvasive testing method, the use of body fluids (such as urine or blood) for the detection of prions offers the best approach.

8.3   Current status of TSE diagnosis in humans

At present there is no accurate diagnosis for TSEs [Brown et al., 2001; Collins et al., 2000; Ingrosso et al., 2002]. For human diseases, diagnosis is based mainly on clinical examination and by exclusion of other, more-common causes of neurological impairments. However, the disease is considered possible or probable depending upon the degree in which the clinical symptoms fit to the standard guidelines, and definitive diagnosis can only be made postmortem through brain histological examination [Ingrosso et al., 2002; Kordek, 2000].

The clinical diagnosis of sCJD is currently based upon the combination of rapidly progressive multifocal dementia with pyramidal and extrapyramidal signs, myoclonus, and visual or cerebellar signs, associated with a characteristic periodic electroencephalogram (EEG) [Collins et al., 2000; Ingrosso et al., 2002; Kordek, 2000; Weber et al., 1997]. A key feature for diagnosing sCJD and distinguishing it from Alzheimer's disease and other dementias is the rapid progression of clinical symptoms and short disease duration, which is often less than 2 years. The clinical manifestation of familial CJD (fCJD) is very similar, except that the disease onset is slightly earlier than in sCJD. Family history of inherited CJD or genetic screening for mutations in the PrP gene are used to establish fCJD diagnosis [Kordek, 2000].

Variant CJD initially appears as a progressive neuropsychiatric disorder characterized by symptoms of anxiety, depression, apathy, withdrawal, and delusions [Henry and Knight, 2002]. Variant CJD is differentiated from sCJD by the duration of illness (usually longer than 6 months) and by the biochemical and histological features of the disease. In addition, because many cases of vCJD have been shown to test positive for PrP^{Sc}-staining in lymphoid tissue (such as tonsil and appendix), a tonsil biopsy is also recommended for vCJD diagnosis [Hill et al., 1999].

Gerstmann-Straussler-Scheinker (GSS) is a predominantly inherited illness that is characterized by dementia, parkinsonian symptoms, and a relatively long duration (typically 5 to 8 years) [Boellaard et al., 1999; Ghetti et al., 1995]. Clinically speaking, GSS is similar to Alzheimer's disease, except that it is often accompanied by ataxia and seizures. Diagnosis is established by clinical examination and genetic screening for PrP mutations [Ghetti et al., 1995]. Fatal familial insomnia (FFI) is also dominantly inherited and associated with PrP mutations. However, the major clinical finding associated with FFI is insomnia, followed at late stages by myoclonus, hallucinations, ataxia, and dementia [Cortelli et al., 1999].

Several biochemical markers have been proposed to be specifically altered in the biological fluids of those afflicted with TSEs, including the proteins S-100, neuron-specific isoenzyme, ubiquitin, the neuronal death marker 14-3-3, and the erythroid differentiation-related factor. However, the only one that is currently used to support clinical diagnosis of sCJD is the measurement of the 14-3-3 protein in the cerebrospinal fluid (CSF) of patients.

8.4   Postmortem detection of BSE in cattle

Several biochemical tests have been developed to rapidly detect animals infected with BSE to prevent the entry of this meat into the food chain. These tests have been evaluated and validated by the European Community [Bird, 2003; Butler, 1998]. Postmortem identification of sick cattle is accurate through histological analysis of the brain [Heim and Wilesmith, 2000]. However, this procedure is time consuming and labor intensive, and it cannot be done on a large scale. The aim of the new rapid tests was to enable processing of many samples, with results being available in just a few hours, so that commercialization of the animals could be withheld until results were available. Based on the results obtained in two blind evaluations, the five tests described in the following subsections were approved by the European Community for BSE detection [Moynagh and Schimmel, 1999b; Schimmel et al., 2002]. Another group of tests is currently under evaluation, and the results will be communicated in the near future. All of the approved tests are based on immunodetection of the pathological PrP^{Sc} isoform either by Western blot or ELISA (enzyme-linked immunosorbent assay) (Figure 8.2), and four of the methods use proteolysis to distinguish PrP^C from PrP^{Sc}.

> *Prionics check, Western blot test:* This test is based on a Western blotting procedure designed to detect the protease-resistant fragment of PrP^{Sc} (denominated as PrP 27–30) [Oesch et al., 2000; Schaller et al., 1999] (Figure 8.2). The minimum time needed to complete the test is about 6 to 8 hours, and approximately 100 tests with duplicates can be performed per person per day. The results of the evaluation of this test showed 100% sensitivity and specificity [Moynagh and Schimmel, 1999a]. When the sensitivity threshold was analyzed in diluted samples, this test scored 15 of the 20 positive samples of the 10^{-1} dilution as positive, 2 as doubtful, and 3 as negative [Moynagh and Schimmel, 1999a]. At the $10^{-1.5}$ dilution, 3 of the 20 positive samples were scored as doubtful, and the remainder were scored as negative.

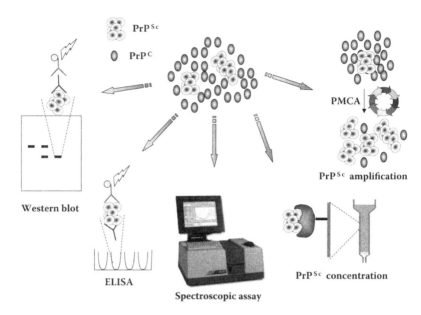

Figure 8.2 Principles of various tests for detection of PrP^{Sc}. The standard tests approved to detect PrP^{Sc} in the brain of cattle rely on the use of antibodies to detect the pathological protein. A mixture of PrP^{C} and PrP^{Sc} is usually subjected to proteolysis to remove the normal protein, and the disease-associated isoform can then be detected by immunological techniques, including Western blots and ELISA. Some of the new procedures under development, such as assays using sophisticated spectroscopic methods or reagents that specifically precipitate PrP^{Sc}, do not rely on proteolysis to distinguish the two isoforms and also do not use antibodies for detection. An interesting alternative to overcome the problem of the very low concentrations of PrP^{Sc} in some samples is to amplify the misfolded protein, using for example the PMCA technology or the cellular infectivity model.

> Below this dilution, this method cannot detect positive samples [Moynagh and Schimmel, 1999a]. The major strength of this method is the good reproducibility and the low rate of false positives; the main disadvantage is the low throughput and relatively low sensitivity compared with other tests.
>
> *ENFER test:* This test is a high-throughput chemiluminescent ELISA (Figure 8.2) that can be completed in less than 4 hours [Moynagh and Schimmel, 1999a; Moynagh and Schimmel, 1999b]. During the European Community evaluation, the ENFER test correctly detected all positive and negative samples, thus resulting in 100% sensitivity and specificity [Moynagh and Schimmel, 1999a]. In the dilution series, this test correctly detected all 10^{-1} and $10^{-1.5}$ but none of the 10^{-2} diluted samples. The main strength of the ENFER test is that the procedure is very rapid and simple; the major disadvantage is the potential for false positives.

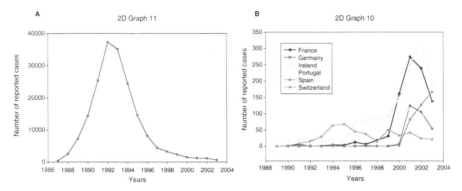

Color Figure 1.1 Cases of bovine spongiform encephalopathy reported in the U.K. (A) and some other European countries (B). (Information was obtained from the Web page of the European Intergovernmental Organization [http://oie.int/info/en_esbru. htm].)

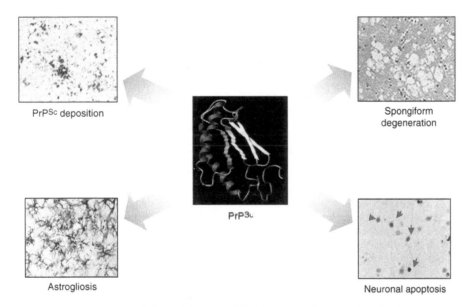

Color Figure 1.2 Neuropathological features of TSE. The central process in prion diseases appears to be the formation of the misfolded prion protein (PrP^{Sc}) that, in some cases, aggregates to form cerebral amyloid deposits. Through a mechanism that is not entirely known, this process induces neuronal apoptosis, spongiform brain degeneration, and astrogliosis.

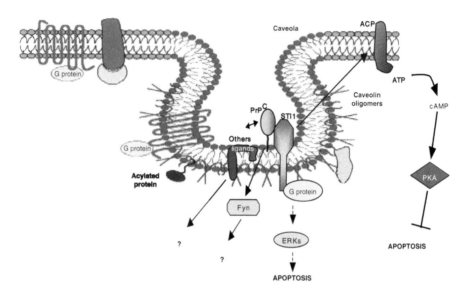

Color Figure 4.1 Signal transduction through the prion protein. Under physiological conditions, PrP^C is located in membrane-specific domains known as lipid rafts or caveola. These membrane structures are rich in receptors and proteins involved in signaling. Interaction between PrP^C and one of its putative ligands (STI1) leads to the potential activation of several signaling pathways: (i) activation of a putative G-protein (Gp), which in turn can activate an adenilate cyclase (ACP), leading to the formation of cAMP and the consequent activation of PKA, and (ii) activation of the ERKs pathway promoting neuronal death. In addition, activation of the tyrosine kinase Fyn by PrP^C via a not-yet-understood mechanism may lead to a signaling pathway with unknown consequences. Finally, interaction of PrP^C with other ligands (laminin receptor, bcl2, neural cell-adhesion molecules, etc.) may also trigger some signaling events.

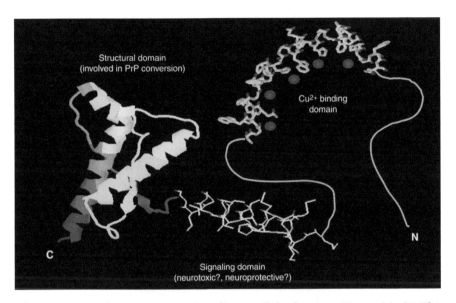

Color Figure 4.2 PrP^C domains involved in diverse cellular functions. The model of PrP^C is based on the known NMR structure of the protein, which indicates the protein regions associated with PrP^C activity. In purple is shown the amino-terminal region in which the five octapeptide repeats are located. This domain has been associated with copper binding, SOD activity, protection from oxidative stress, and Bax-induced apoptosis. In red is shown the neurotoxic domain from amino acids 106–126, which has been involved in the interaction with STI1 protein and the initiation of PKA and ERK signaling. The carboxy-terminal globular domain is implicated in the conversion of PrP^C into PrP^{Sc}.

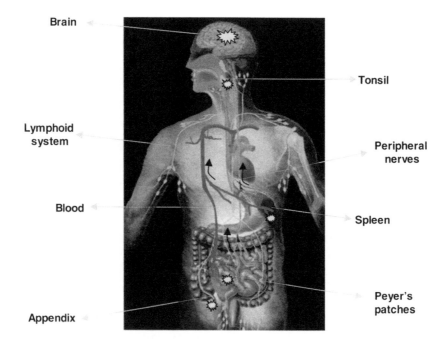

Brain

Tonsil

Lymphoid system

Peripheral nerves

Blood

Spleen

Peyer's patches

Appendix

☼ **Sites of prion replication and/or accumulation**

↑ **Sites of prion transport**

Color Figure 6.1 Schematic representation of the pathways implicated in the transport and replication of prions. Orally ingested prions are intestinally absorbed mainly at the level of Peyer's patches and transported to the blood and lymphoid fluids. After a peripheral replication in spleen, appendix, tonsil, and other lymphoid tissues, prions are transported to the brain mainly via peripheral nerves. A direct penetration to the brain across the blood–brain barrier is also possible.

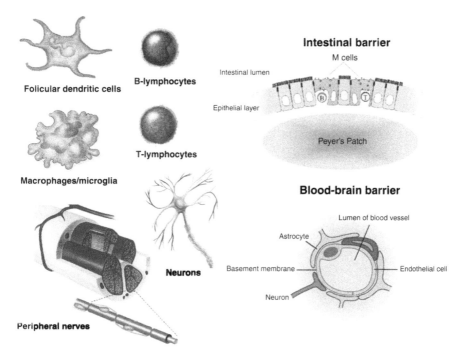

Color Figure 6.2 Cells implicated in prion diseases. Diverse cell types have been implicated in the transport of prions (B- and T-lymphocytes, peripheral nerves, M cells), in the replication of the infectious agent (neurons, follicular dendritic cells), and in the clearance of prions (macrophages, microglia). Also, to reach the brain, the infectious agent has to cross at least two tight barriers, the intestinal and the blood–brain barrier.

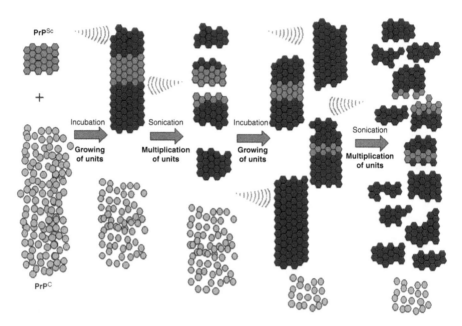

Color Figure 10.1 Diagrammatic representation of PMCA. Cyclic amplification consists of subjecting a sample containing minute quantities of PrP^{Sc} and a large excess of PrP^C to cycles of incubation/sonication. During incubation, PrP^{Sc} aggregates grow by converting and incorporating PrP^C into the polymer. Sonication is used to break down large aggregates into many smaller pieces to multiply the number of converting units. PMCA cycles can be repeated as many times as needed to amplify undetectable quantities of PrP^{Sc} to a level of easy detection.

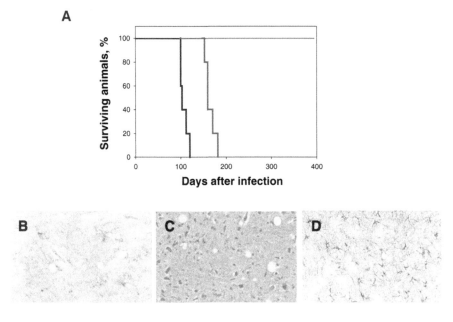

Color Figure 10.2 Infectious properties of *in vitro*-generated PrP^{Sc}: (A) Wild-type hamsters, inoculated with similar quantities of PrP^{Sc} either generated *in vitro* or derived from the brain, developed the disease. The graph shows the survival time for each animal in groups inoculated with each brain infectious material or with *in vitro*-generated PrP^{Sc} under conditions in which no molecules of inoculum PrP^{Sc} are present. (B) PrP^{Sc} accumulates in the brain of animals sick after inoculation with *in vitro*-generated infectivity. (C) Hematoxilin-eosin staining of the brain revealed vacuolation, which represents the typical spongiform degeneration observed in TSEs. (D) Astrocytes, another of the typical brain alterations observed in these animals, were studied by glial fibrillary acidic protein (GFAP) staining.

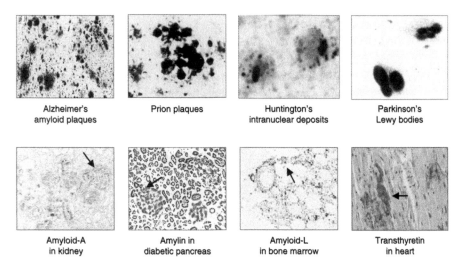

Color Figure 11.1 Histological staining of protein deposits in different tissues. Several distinct proteins have the ability to misfold and accumulate as amyloid-like plaques in different organs, triggering tissue damage and organ dysfunction.

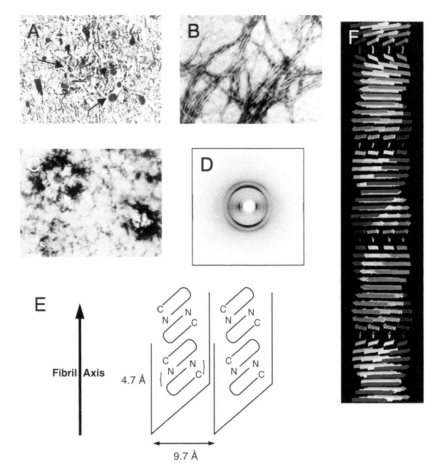

Color Figure 11.2 Morphological, tinctorial, and structural characteristics of amyloid aggregates. (A) A tissue staining with antibodies specific for the protein forming the aggregates reveals the accumulation of many punctuated aggregates. The localization of this protein deposit differs depending on the protein component and the particular disease. (B) Electron microscopy shows typical unbranched fibrils, which are 5 to 10 nm in diameter and 100 nm in length. (C) Cong red staining visualized under polarized light shows green/yellow birefringency, which is considered to be a typical signature for amyloid aggregates. (D) X-ray fiber diffraction studies show a characteristic pattern known as cross-β, represented by signals at 4.7 and 9.7 Å. (E) A diagrammatic picture for the cross-β structure showing the periodicity observed in X-ray diffraction studies. (F) A computer model of the putative tridimensional structure of a prototype amyloid fibril.

CEA/BioRad test: This test is a sandwich immunoassay (Figure 8.2) for PrP 27–30 carried out after denaturation and concentration steps [Grassi et al., 2001; Moynagh and Schimmel, 1999a]. Two different monoclonal antibodies are used. As with the previous tests, the CEA/BioRad assay obtained 100% sensitivity and specificity during the evaluation [Moynagh and Schimmel, 1999a]. In the dilution assay, this test correctly identified all dilutions down to 10^{-2} and even 18 out of 20 from $10^{-2.5}$ diluted samples. The major strength of this test is its high sensitivity and the demonstration that it can detect PrP^{Sc} in the brain of presymptomatic animals. The main weaknesses are the more time-consuming and laborious procedure and the potential for false positives.

Prionics check, LIA test: This assay is also a sandwich ELISA (Figure 8.2) using two different monoclonal antibodies to detect PK-resistant PrP [Biffiger et al., 2002]. The system is made for a 96-well format and can be fully automated from homogenization to detection and analysis. Processing time is less than 4 hours. Sensitivity was 97.9% and specificity was 100% in the European evaluation [Schimmel et al., 2002]. In the dilution study, this test correctly identified the samples up to 10^{-1} or 10^{-2} dilution, depending upon the experimental conditions used for protease digestion and tissue homogenization. The major advantage is that the system is fairly simple, rapid, high throughput, and automated; the main disadvantage is the reported variability depending on sample preparation and the potential for false positives.

Conformational dependent immunoassay (CDI) test: This test consists of a sandwich immunoassay with time-resolved fluorescence detection employing high-affinity recombinant antibody fragments for the detection and capture of PrP [Safar et al., 1998; Safar et al., 2002]. The CDI simultaneously measures specific antibody binding to denatured and native forms of PrP. The detection antibody recognizes a conformation-dependent epitope, which while always exposed in PrP^{C}, only becomes exposed in PrP^{Sc} upon denaturation [Safar et al., 1998]. Unlike other assays used for rapid detection of prions, the CDI does not depend on the protease-resistant properties of PrP^{Sc}. In its current state of development, this test takes less than 8 hours to complete. The results of the European evaluation of this test showed a 100% sensitivity and specificity. In the dilution series study, out of the four samples analyzed for dilution $10^{-1.5}$, two were clearly positive and two were borderline positive. For dilution 10^{-2}, one positive and two borderline positive results were obtained [Schimmel et al., 2002]. The major strength of this method is its independence of protease digestion, which enables detection of protease-sensitive versions of PrP^{Sc}. The main weakness is that the procedure is more cumbersome, involving several steps.

8.5   The need for detection of PrP^{Sc} in blood

The development of the rapid BSE tests has been important because it decreases the risk of sick animals entering the food chain. However, based on the estimated sensitivities, it is unlikely that, in their current status, they could be used to detect animals at early stages of the infectious period [Schiermeier, 2001]. It is also unlikely that they can detect PrP^{Sc} in tissues other than brain. The latter is essential for the development of noninvasive tests that can be used in living individuals.

The possibility of detecting PrP^{Sc} in peripheral tissues has been clearly demonstrated in the case of vCJD, in which the misfolded protein can be detected in lymphoid tissue [Hill et al., 1997; Hill et al., 1999]. Interestingly, there have been reports of individual cases showing detection of PrP^{Sc} at preclinical stages of the disease in tonsil [Schreuder et al., 1996] as well as in the appendix [Hilton et al., 1998], indicating that lymphoid tissue biopsy may be useful for diagnosing presymptomatic individuals. Although PrP^{Sc}-positive lymphoid tissue was considered to be specific for vCJD [Hill et al., 1999; Wadsworth et al., 2001], recent studies identified PrP^{Sc} in as many as one-third of skeletal muscle and spleen samples [Glatzel et al., 2003] as well as in the olfactory epithelium [Zanusso et al., 2003] of patients suffering from sCJD.

Routine laboratory diagnosis of any disease in presymptomatic individuals is limited to the examination of external tissues or fluids. Detection of a protease-resistant PrP^{res}-like form in the urine of humans and animals affected by TSE has been reported [Shaked et al., 2001]. However, recent studies have questioned the interpretation of these results, suggesting that the PrP^{Sc}-like reactivity detected in urine corresponded to contamination with bacterial proteins [Furukawa et al., 2004] or to immunoglobulins detected by the secondary antibody [Serban et al., 2004]. Nevertheless, neither urine nor any of the other external tissues or peripheral fluids (with the exception of blood) has ever been shown to be infectious in human beings or experimental animals with TSE [Brown et al., 2001]. Hence, blood seems the best option for a routinely noninvasive diagnosis of TSE.

From infectivity studies in rodent models of TSE, it seems that the maximum concentration of infectivity in circulating blood resides in the buffy coat, where it may be present at a level of 5 to 10 infectious units (IU) per mL during the incubation period and then rise to around 100 IU/mL at the onset of symptomatic disease [Brown et al., 1998; Brown et al., 2001; Brown, 2001; Holada et al., 2002]. As little as 1 pg of PrP^{Sc} is estimated to contain 100 IU [Brown et al., 2001]; hence, the concentration of PrP^{Sc} in the buffy coat is expected to be 1 pg/mL ($\approx 3 \times 10^{-14}M$) and 0.1 pg/mL ($\approx 3 \times 10^{-15}M$) at the symptomatic and presymptomatic phases, respectively. The concentrations of PrP^{Sc} in plasma and red blood cells are estimated to be at least five- to ten-fold lower than in the buffy coat [Brown et al., 1998]. Therefore, for a test to be able to detect PrP^{Sc} in blood samples during the presymptomatic phase of the disease, it would have to be able to detect the protein in the femtomolar concentration range. In other words, such a test would have to be able to detect as few as 60,000 molecules of PrP^{Sc}, which is the amount estimated to be present in 100 µL of rodent blood. However, based on the comparison of the infectivity in brain, it is likely that in human and cattle blood samples, PrP^{Sc} concentrations are even lower.

8.6   Novel approaches under development for premortem early diagnosis

Several groups are currently attempting to develop new strategies with enough sensitivity to detect PrP^{Sc} in blood [Soto, 2004]. Following is a brief description of the different technologies under development, their principles, and their major strengths and weaknesses.

8.6.1   Spectroscopic techniques

Several sophisticated techniques based on the use of spectrophotometry (Figure 8.2) are under development with the aim of creating a high-sensitivity detection test for PrP^{res}.

Multispectral ultraviolet fluoroscopy: Multispectral ultraviolet fluoroscopy (MUFS) identifies proteins by their specific fluorescent pattern of emission when they are excited in the ultraviolet range [Rubenstein et al., 1998]. This technique has been shown to discriminate between PrP^{C} and PrP^{Sc} as well as various forms of PrP^{Sc} from different strains. The main strength of this method is that it does not rely on antibodies or protease digestion for PrP^{Sc} detection, thus providing an alternative to the other tests. However, in its current stage, the sensitivity of this assay is still lower than ELISA or Western blot tests.

Confocal dual-color fluorescence-correlation spectroscopy: Confocal dual-color fluorescence-correlation spectroscopy (FCS) recognizes single fluorescent molecules in solution when they pass through the exciting laser beam in a confocal microscope equipped with a single-photon counter [Bieschke et al., 2000]. In this method, the sample is mixed with anti-PrP antibodies labeled with two different fluorescent dies. PrP^{Sc} aggregates offer multiple sites for antibody binding, producing intensely fluorescent molecules that can be scanned and separated from monomeric PrP^{C}. This technique appears to be more sensitive than Western blots and ELISA assays and was shown to be able to detect PrP^{Sc} in the CSF of a subset of sCJD patients but not in normal controls [Bieschke et al., 2000]. However, the massive application of FCS for TSE diagnosis does not seem feasible because of the requirement of expensive and sophisticated equipment that is not readily available.

FTIR spectra analyzed by neural network: Analysis of Fourier-transformed infrared spectroscopy (FTIR) spectra of samples containing PrP^{Sc} by artificial neural networks has been reported to produce an assay sensitive enough to diagnose TSE using blood samples [Lasch et al., 2003; Schmitt et al., 2002]. The test consists of measuring IR spectra of many control and prion-infected samples that, after derivatization and transformation of the spectra, are used as an input to the analysis of the artificial neural network [Kneipp et al., 2000; Lasch et al., 2003]. In a training phase, the software finds patterns to distinguish samples from normal and sick animals. After the training and validation phases, the software is ready to analyze unknown samples and classify them as coming from either normal or infected animals. Using this analysis, it was shown that the test gave 97% sensitivity and 100% specificity in the identification of normal and diseased cases using blood samples [Lasch et al., 2003]. The technology has also been used to detect infected animals in the presymptomatic phase of the disease [Kneipp et al., 2002]. These intriguing findings are promising for the development of an accurate blood test for TSE. The major weakness of this assay is that it is not clear what the method is really detecting, since the analysis is aimed at identifying differences in the FTIR spectra of samples coming from normal and infected individuals. In addition, the method requires very sophisticated skills to measure and interpret the results, and thus its widespread application for blood diagnosis might be difficult to implement.

Fluorescence detection after capillary electrophoresis: A method based on fluorescence detection after capillary electrophoresis (CE) in combination with antigen/antibody competition has been reported by Schmerr and colleagues [Jackman and Schmerr, 2003; Schmerr et al., 1999; Schmerr and Jenny, 1998]. In this assay, a fluorescein-labeled synthetic peptide bearing a short PrP sequence is incubated with an anti-PrP antibody in a proportion carefully estimated to produce binding to 50% of the peptide. This material is mixed with the sample treated with PK, and if PrP^{Sc} is present, it will displace the peptide from the antibody, changing the proportion of free and bound peptide [Jackman and Schmerr, 2003]. This proportion is then measured by laser-induced fluorescence during capillary-zone electrophoresis. This method has been reported to detect PrP^{Sc} in the blood of scrapie-affected sheep and elks with chronic wasting disease [Schmerr et al., 1999]. The major weaknesses of this technology are that it seems difficult to standardize, and it does not directly measure PrP^{Sc} but, rather, depends upon the comparison of signal ratios among different samples. In addition, these results have not been reproduced in other laboratories [Cervenakova et al., 2003].

8.6.2   Conformational antibodies

Production of PrP^{Sc}-specific conformational antibodies may be very helpful for diagnosis, helping to detect quantities of PrP^{Sc} in fluids and enabling the concentration of specific markers (Figure 8.2). Considering the large structural differences between PrP^{C} and PrP^{Sc}, it was thought that production of conformational antibodies could be a feasible task. However, attempts by many different groups using several alternative strategies have systematically failed [Demart et al., 1999; Groschup et al., 1997; Kascsak et al., 1997]. The first successful result was reported in 1997 for the antibody 15B3, which specifically immunoprecipitates bovine, murine, or human PrP^{Sc} but not PrP^{C}, suggesting that it recognizes an epitope common to prions from different species [Korth et al., 1997; Korth et al., 1999]. However, the practical use of this antibody for prion diagnosis has not been pursued, most likely because of its relatively low affinity for the protein.

More recently, three other putative PrP^{Sc}-specific antibodies have been reported. One of the new strategies used to generate conformational antibodies is based on experiments showing that induction of β-sheet structures in recombinant PrP is associated with increased solvent accessibility of tyrosine residues [Paramithiotis et al., 2003]. Because tyrosine is usually found in the tyrosine-tyrosine-arginine repeat motif, animals were immunized with synthetic peptides bearing this sequence. Several of the antibodies produced recognized PrP^{Sc} (but not PrP^{C}), as assessed by immunoprecipitation, plate capture immunoassay, and flow cytometry [Paramithiotis et al., 2003]. Another reported PrP^{Sc}-specific monoclonal antibody, named V5B2, was generated against a C-terminal PrP synthetic peptide [Curin et al., 2004]. This antibody recognized PrP^{Sc} (but not PrP^{C}) from humans affected by CJD using dot blot, ELISA, immunoprecipitation, and immunohistochemistry. Finally, it was recently published that an anti-DNA antibody, named OCD4, captured PrP from brains affected by prion diseases in both humans and animals but not from unaffected controls [Zou et al., 2004]. OCD4 appears to immunoreact with DNA (or a DNA-associated molecule) that forms a conformation-dependent complex with PrP in prion diseases. Whereas PrP immunocaptured by OCD4 is largely protease resistant, a fraction of it remains protease sensitive.

Further studies are needed to analyze the specificity and usefulness of these three conformational antibodies for TSE diagnosis. One important note of caution to the identification of conformational antibodies was recently reported by Morel et al. in a study in which several PrP-directed antibodies as well as totally unrelated antibodies were shown to immunoprecipitate PrP^{Sc} but not PrP^{C} [Morel et al., 2004]. The authors concluded that binding and precipitation of PrP^{Sc} by antibodies depend upon nonspecific interactions.

8.6.3 PrP^Sc concentration by binding to specific ligands

A strategy under study to increase the sensitivity of PrP^Sc detection is to specifically enrich the marker in the sample prior to detection (Figure 8.2). Several groups are attempting to identify ligands able to bind specifically to PrP^Sc. Such molecules would be useful to precipitate minute quantities of the marker, enabling concentration and detection. They might also be used in place of conformational antibodies in ELISA or dot-blot assays. It has been reported that sodium phosphotungstic acid selectively precipitates PrP^Sc from different sources [Safar et al., 1998; Wadsworth et al., 2001]. This procedure, coupled with an enhanced chemiluminescence detection system, has led to the development of a highly sensitive test for identifying PrP^Sc in diverse samples. Using this assay, Collinge and coworkers have detected PrP^Sc in various tissues from vCJD patients, including spinal cord, thymus, lymph nodes, tonsil, spleen, adrenal gland, rectum, retina, and optic nerve [Wadsworth et al., 2001]. However, detection of PrP^Sc in blood samples was not possible using this test.

Another molecule reported to interact specifically with PrP^Sc is plasminogen obtained from human and mouse blood [Fischer et al., 2000]. Furthermore, plasminogen seems to bind PrP^Sc from multiple species [Maissen et al., 2001]. However, it has been reported by other groups that plasminogen also exhibits a remarkable affinity for PrP^C [Kornblatt et al., 2003; Praus et al., 2003; Shaked et al., 2002], which may compromise its application for diagnosis. Another strategy that has been analyzed takes advantage of the known capacity of PrP^Sc to bind nucleic acids, in particular RNA [Akowitz et al., 1994; Deleault et al., 2003; Gabus et al., 2001; Weiss et al., 1997]. Artificial ligands or aptamers have been identified by *in vitro* selection from a large randomized library of 2'-fluoro-modified RNA [Rhie et al., 2003]. One of these ligands (aptamer SAF-93) was found to have more than ten-fold higher affinity for PrP^Sc than for recombinant PrP^C and to have inhibited the accumulation of PrP^Sc in a cell-free conversion assay [Proske et al., 2002; Sayer et al., 2004]. The application of this and other RNA aptamers for prion diagnosis remains to be studied.

8.6.4 PrP^Sc amplification

An interesting alternative to increasing the sensitivity of PrP^Sc detection is to amplify the amount of the marker present in a sample (Figure 8.2). With this purpose in mind, we developed a technology called protein misfolding cyclic amplification (PMCA), which is based on mimicking the replication of prions *in vivo* [Saborio et al., 2001]. At the time of infection, animals or humans get infected with minute and undetectable amounts of PrP^Sc, which replicate at the expense of PrP^C such that, by the time the clinical symptoms appear, the brain is full of the abnormal protein. This replication process is slow; indeed, in humans it can take up to 40 years for PrP^Sc to grow enough to trigger the disease. The PMCA procedure successfully mimics prion replication in the test tube, but in an accelerated manner. (For a detailed description of the principles behind PMCA and its applications, see Chapter 10.) In our original publication, we showed that PMCA was able to increase the sensitivity of detection by 30- to 50-fold [Saborio et al., 2001], but in our current optimized and automated procedure, the sensitivity of detection is over 10-million-fold higher than existing tests, making it possible to detect the small quantities of PrP^{Sc} present in blood [Castilla et al., 2005].

Another strategy employed to amplify PrP^{Sc} is the use of cell infectivity assays [Klohn et al., 2003]. This assay utilizes mouse neuroblastoma N2a sublines that are highly susceptible to infection with mouse prions [Bosque and Prusiner, 2000; Enari et al., 2001]. In this assay, susceptible N2a cells are exposed to prion-containing samples for 3 days, grown to confluence, and split three times. The dose response to infection is linear over two logs of prion concentrations. The cell assay was claimed to be as sensitive as the mouse bioassay, ten times faster, two orders of magnitude less expensive, and suitable for robotization [Klohn et al., 2003]. A potential problem with this assay is that, up to now, there have not been cellular systems available for propagating prions from the most relevant cattle and human origins. The applicability of this promising approach for routine TSE diagnosis needs to be studied further.

8.7   Concluding remarks

The unprecedented nature of the prion infectious material and the complicated mechanism of transmission have made the development of diagnosis procedures very challenging. However, over the past 5 years, great progress has been made in TSE diagnosis, and several of the new strategies hold substantial promise for early, noninvasive detection of infected, but still clinically unaffected, individuals. In addition, the parallel increase of knowledge of the molecular basis of the disease may lead to the design of novel approaches for diagnosis or the identification of new specific markers for the disease.

References* **

Aguzzi, A., Prion diseases, blood and the immune system: concerns and reality, *Haematologica*, 85, 3–10, 2000.

Aguzzi, A. et al., Interventional strategies against prion diseases. *Nat. Rev. Neurosci.*, 2, 745–749, 2001.

Akowitz, A., Sklaviadis, T., and Manuelidis, L., Endogenous viral complexes with long RNA cosediment with the agent of Creutzfeldt-Jakob disease, *Nucleic Acids Res.*, 22, 1101–1107, 1994.

---

* Highlights primary articles of outstanding importance and quality, including a short description of the findings.
** Highlights comprehensive review articles similar to the topic of this chapter.

Bieschke, J. et al., Ultrasensitive detection of pathological prion protein aggregates by dual-color scanning for intensely fluorescent targets, *Proc. Natl. Acad. Sci. USA*, 97, 5468–5473, 2000.

Biffiger, K. et al., Validation of a luminescence immunoassay for the detection of PrP(Sc) in brain homogenate, *J. Virol. Methods*, 101, 79–84, 2002.

Bird, S.M., European Union's rapid TSE testing in adult cattle and sheep: implementation and results in 2001 and 2002, *Stat. Methods Med. Res.*, 12, 261–278, 2003.

Boellaard, J.W., Brown, P., and Tateishi, J., Gerstmann-Straussler-Scheinker disease: the dilemma of molecular and clinical correlations, *Clin. Neuropathol.*, 18, 271–285, 1999.

Bosque, P.J. and Prusiner, S.B., Cultured cell sublines highly susceptible to prion infection, *J. Virol.*, 74, 4377–4386, 2000.

Brown, P., The risk of blood-borne Creutzfeldt-Jakob disease, *Dev. Biol. Stand.*, 102, 53–59, 2000.

*Brown, P., Creutzfeldt-Jakob disease: blood infectivity and screening tests, *Semin. Hematol.*, 38, 2–6, 2001.

*Brown, P., Cervenakova, L., and Diringer, H., Blood infectivity and the prospects for a diagnostic screening test in Creutzfeldt-Jakob disease, *J. Lab Clin. Med.*, 137, 5–13, 2001.

Brown, P. et al., Iatrogenic Creutzfeldt-Jakob disease at the millennium, *Neurology*, 55, 1075–1081, 2000.

Brown, P. et al., The distribution of infectivity in blood components and plasma derivatives in experimental models of transmissible spongiform encephalopathy, *Transfusion*, 38, 810–816, 1998.

Butler, D., Brussels seeks BSE diagnostic test to screen European cattle, *Nature*, 395, 205–206, 1998.

Castilla, J., Saa, P., and Soto, C., Biochemical detection of prions in blood, *Nature Med.*, 11, 982–985, 2005 (Reports for the first time the detection of PrP^{Sc} in blood).

Cervenakova, L. et al., Failure of immunocompetitive capillary electrophoresis assay to detect disease-specific prion protein in buffy coat from humans and chimpanzees with Creutzfeldt-Jakob disease, *Electrophoresis*, 24, 853–859, 2003.

Collinge, J., Variant Creutzfeldt-Jakob disease, *Lancet*, 354, 317–323, 1999.

Collins, S. et al., Recent advances in the pre-mortem diagnosis of Creutzfeldt-Jakob disease, *J. Clin. Neurosci.*, 7, 195–202, 2000.

Cortelli, P. et al., Fatal familial insomnia: clinical features and molecular genetics, *J. Sleep Res.*, 8 (supp. 1), 23–29, 1999.

Curin, S. et al., Monoclonal antibody against a peptide of human prion protein discriminates between Creutzfeldt-Jacob's disease–affected and normal brain tissue, *J. Biol. Chem.*, 279, 3694–3698, 2004.

Deleault, N.R., Lucassen, R.W., and Supattapone, S., RNA molecules stimulate prion protein conversion, *Nature*, 425, 717–720, 2003.

Demart, S. et al., New insight into abnormal prion protein using monoclonal antibodies, *Biochem. Biophys. Res. Commun.*, 265, 652–657, 1999.

Donnelly, C.A. et al., Extending backcalculation to analyse BSE data, *Stat. Methods Med. Res.*, 12, 177–190, 2003.

Dormont, D., Approaches to prophylaxis and therapy, *Br. Med. Bull.*, 66, 281–292, 2003.

Enari, M., Flechsig, E., and Weissmann, C., Scrapie prion protein accumulation by scrapie-infected neuroblastoma cells abrogated by exposure to a prion protein antibody, *Proc. Natl. Acad. Sci. USA*, 98, 9295–9299, 2001.

Field, E.J. and Shenton, B.K., Rapid diagnosis of scrapie in the mouse, *Nature*, 240, 104–106, 1972.

Fischer, M.B. et al., Binding of disease-associated prion protein to plasminogen, *Nature*, 408, 479–483, 2000.

Furukawa, H. et al., A pitfall in diagnosis of human prion diseases using detection of protease-resistant prion protein in urine: contamination with bacterial outer membrane proteins, *J. Biol. Chem.*, 279, 23661–23667, 2004.

Gabus, C. et al., The prion protein has RNA binding and chaperoning properties characteristic of nucleocapsid protein NCP7 of HIV-1, *J. Biol. Chem.*, 276, 19301–19309, 2001.

Ghetti, B. et al., Gerstmann-Straussler-Scheinker disease and the Indiana kindred, *Brain Pathol.*, 5, 61–75, 1995.

'Glatzel, M. et al., Extraneural pathologic prion protein in sporadic Creutzfeldt-Jakob disease, *N. Engl. J. Med.*, 349, 1812–1820, 2003. (Reports the detection of PrP^{Sc} in several peripheral tissues of patients affected by CJD.)

Grassi, J. et al., Rapid test for the preclinical postmortem diagnosis of BSE in central nervous system tissue, *Vet. Rec.*, 149, 577–582, 2001.

Groschup, M.H., Harmeyer, S., and Pfaff, E., Antigenic features of prion proteins of sheep and of other mammalian species, *J. Immunol. Methods*, 207, 89–101, 1997.

Heim, D. and Wilesmith, J.W., Surveillance of BSE, *Arch. Virol. Suppl.*, 16, 127–133, 2000.

Henry, C. and Knight, R., Clinical features of variant Creutzfeldt-Jakob disease, *Rev. Med. Virol.*, 12, 143–150, 2002.

'Hill, A.F. et al., Investigation of variant Creutzfeldt-Jakob disease and other human prion diseases with tonsil biopsy samples, *Lancet*, 353, 183–189, 1999. (This and the following two articles report important results on the feasibility of detecting PrP^{Sc} in tonsil and appendix samples before the onset of clinical symptoms.)

'Hill, A.F. et al., Diagnosis of new variant Creutzfeldt-Jakob disease by tonsil biopsy, *Lancet*, 349, 99–100, 1997.

'Hilton, D.A. et al., Prion immunoreactivity in appendix before clinical onset of variant Creutzfeldt Jakob disease, *Lancet*, 352, 703–704, 1998.

Holada, K. et al., Scrapie infectivity in hamster blood is not associated with platelets, *J. Virol.*, 76, 4649–4650, 2002.

''Ingrosso, L. et al., Molecular diagnostics of transmissible spongiform encephalopathies, *Trends Mol. Med.*, 8, 273–280, 2002.

Jackman, R. and Schmerr, M.J., Analysis of the performance of antibody capture methods using fluorescent peptides with capillary zone electrophoresis with laser-induced fluorescence, *Electrophoresis*, 24, 892–896, 2003.

Kascsak, R.J. et al., Immunodiagnosis of prion disease, *Immunol. Invest.*, 26, 259–268, 1997.

'Klohn, P.C. et al., A quantitative, highly sensitive cell-based infectivity assay for mouse scrapie prions, *Proc. Natl. Acad. Sci. USA*, 100, 11666–11671, 2003. (Describes the use of cell-infectivity studies to amplify the PrP^{Sc} signal, increasing sensitivity of detection.)

Kneipp, J. et al., Molecular changes of preclinical scrapie can be detected by infrared spectroscopy, *J. Neurosci.*, 22, 2989–2997, 2002.

Kneipp, J. et al., Detection of pathological molecular alterations in scrapie-infected hamster brain by Fourier transform infrared (FT-IR) spectroscopy, *Biochim. Biophys. Acta*, 1501, 189–199, 2000.

**Kordek, R., The diagnosis of human prion diseases, *Folia Neuropathol.*, 38, 151–160, 2000.

Kornblatt, J.A. et al., The fate of the prion protein in the prion/plasminogen complex, *Biochem. Biophys. Res. Commun.*, 305, 518–522, 2003.

Korth, C. et al., Prion (PrPSc)-specific epitope defined by a monoclonal antibody, *Nature*, 390, 74–77, 1997.

Korth, C., Streit, P., and Oesch, B., Monoclonal antibodies specific for the native, disease-associated isoform of the prion protein, *Methods Enzymol.*, 309, 106–122, 1999.

Lasch, P. et al., Antemortem identification of bovine spongiform encephalopathy from serum using infrared spectroscopy, *Anal. Chem.*, 75, 6673–6678, 2003.

*Llewelyn, C.A. et al., Possible transmission of variant Creutzfeldt-Jakob disease by blood transfusion, *Lancet*, 363, 417–421, 2004. (A very important study reporting the possible first case of vCJD transmission by blood transfusion.)

Maissen, M. et al., Plasminogen binds to disease-associated prion protein of multiple species, *Lancet*, 357, 2026–2028, 2001.

Morel, N. et al., Selective and efficient immunoprecipitation of the disease-associated form of the prion protein can be mediated by nonspecific interactions between monoclonal antibodies and scrapie-associated fibrils, *J. Biol. Chem.*, 279, 30143–30149, 2004.

*Moynagh, J. and Schimmel, H., Tests for BSE evaluated: bovine spongiform encephalopathy, *Nature*, 400, 105, 1999a. (Reports the results of the first evaluation of BSE rapid tests in Europe.)

Moynagh, J. and Schimmel, H., The Evaluation of Tests for the Diagnosis of Transmissible Spongiform Encephalopathy in Bovines, report from the European Commission, 8 July 1999b; available on-line at http://europa.eu.int/comm/food/fs/bse/bse12_en.pdf.

Nolen, R.S., Washington state dairy cow nation's first case of BSE, *J. Am. Vet. Med. Assoc.*, 224, 345–346, 2004.

Oesch, B. et al., Application of prionics western blotting procedure to screen for BSE in cattle regularly slaughtered at Swiss abattoirs, *Arch. Virol. Suppl.*, 16, 189–195, 2000.

Paramithiotis, E. et al., A prion protein epitope selective for the pathologically misfolded conformation, *Nature Med.*, 9, 893–899, 2003.

Peden, A.H. et al., Preclinical vCJD after blood transfusion in a PRNP codon 129 heterozygous patient, *Lancet*, 364, 527–529, 2004.

Praus, M. et al., Stimulation of plasminogen activation by recombinant cellular prion protein is conserved in the NH2-terminal fragment PrP23-110, *Thromb. Haemost.*, 89, 812–819, 2003.

Proske, D. et al., Prion-protein-specific aptamer reduces PrPSc formation, *Chembiochem.*, 3, 717–725, 2002.

Prusiner, S.B., Prions, *Proc. Natl. Acad. Sci. USA*, 95, 13363–13383, 1998.

Rhie, A. et al., Characterization of 2'-fluoro-RNA aptamers that bind preferentially to disease-associated conformations of prion protein and inhibit conversion, *J. Biol. Chem.*, 278, 39697–39705, 2003.

Rossi, G. et al., Therapeutic approaches to prion diseases, *Clin. Lab. Med.*, 23, 187–208, 2003.

Rubenstein, R. et al., Detection and discrimination of PrPSc by multi-spectral ultraviolet fluorescence, *Biochem. Biophys. Res. Commun.*, 246, 100–106, 1998.

*Saborio, G.P., Permanne, B., and Soto, C., Sensitive detection of pathological prion protein by cyclic amplification of protein misfolding, *Nature*, 411, 810–813, 2001. (Describes the amplification of PrP^{Sc} *in vitro*, which may have important applications to increase sensitivity of prion diagnosis.)

Safar, J. et al., Eight prion strains have PrP(Sc) molecules with different conformations, *Nature Med.*, 4, 1157–1165, 1998.

Safar, J.G. et al., Measuring prions causing bovine spongiform encephalopathy or chronic wasting disease by immunoassays and transgenic mice, *Nature Biotechnol.*, 20, 1147–1150, 2002.

Sayer, N.M. et al., Structural determinants of conformationally selective, prion-binding aptamers, *J. Biol. Chem.*, 279, 13102–13109, 2004.

Schaller, O. et al., Validation of a western immunoblotting procedure for bovine PrP(Sc) detection and its use as a rapid surveillance method for the diagnosis of bovine spongiform encephalopathy (BSE), *Acta Neuropathol. (Berl.)*, 98, 437–443, 1999.

Schiermeier, Q., Testing times for BSE, *Nature*, 409, 658–659, 2001.

Schimmel, H. et al., The evaluation of five rapid tests for the diagnosis of transmissible spongiform encephalopathy in bovines (second study), report from the European Commission, 27 March 2002; available on-line at http://europa.eu.int/comm/food/fs/bse/bse42_en.pdf

Schmerr, M.J. and Jenny, A., A diagnostic test for scrapie-infected sheep using a capillary electrophoresis immunoassay with fluorescent-labeled peptides, *Electrophoresis*, 19, 409–414, 1998.

Schmerr, M.J. et al., Use of capillary electrophoresis and fluorescent labeled peptides to detect the abnormal prion protein in the blood of animals that are infected with a transmissible spongiform encephalopathy, *J. Chromatogr. A*, 853, 207–214, 1999.

Schmitt, J. et al., Identification of scrapie infection from blood serum by Fourier transform infrared spectroscopy, *Anal. Chem.*, 74, 3865–3868, 2002.

Schreuder, B.E. et al., Preclinical test for prion diseases, *Nature*, 381, 563, 1996.

Serban, A. et al., Immunoglobulins in urine of hamsters with scrapie, *J. Biol. Chem.*, 279, 18817 18820, 2004.

Shaked, G.M. et al., A protease-resistant prion protein isoform is present in urine of animals and humans affected with prion diseases, *J. Biol. Chem.*, 276, 31479–31482, 2001.

Shaked, Y., Engelstein, R., and Gabizon, R., The binding of prion proteins to serum components is affected by detergent extraction conditions, *J. Neurochem.*, 82, 1–5, 2002.

**Soto, C., Diagnosing prion diseases: needs, challenges and hopes, *Nature Rev. Microbiol.*, 2, 809–819, 2004.

Wadsworth, J.D. et al., Tissue distribution of protease resistant prion protein in variant Creutzfeldt-Jakob disease using a highly sensitive immunoblotting assay. *Lancet*, 358, 171–180, 2001.

Weber, T. et al., Diagnosis of Creutzfeldt-Jakob disease and related human spongiform encephalopathies, *Biomed. Pharmacother.*, 51, 381–387, 1997.

Weiss, S. et al., RNA aptamers specifically interact with the prion protein PrP, *J. Virol.*, 71, 8790–8797, 1997.

Zanusso, G. et al., Detection of pathologic prion protein in the olfactory epithelium in sporadic Creutzfeldt-Jakob disease, *N. Engl. J. Med.*, 348, 711–719, 2003.

Zou, W.Q. et al., Antibody to DNA detects scrapie but not normal prion protein, *Proc. Natl. Acad. Sci. USA*, 101, 1380–1385, 2004.

chapter nine

Therapeutic approaches

Transmissible spongiform encephalopathies (TSEs) are 100% fatal diseases. Currently there is no approved treatment for these diseases. Only a few molecules have been tested in human clinical trials, and so far all of them have failed to produce any significant improvement. The little effort spent in finding therapies stands in stark contrast to the great progress in understanding the molecular basis of the disease pathogenesis. The reason for the slow transition from the scientific advances to clinical applications is likely due to the pharmaceutical industry's lack of interest in these rare diseases. However, in the last few years, many academic groups have been attempting to implement novel therapeutic approaches, and with the help of government funds, some of them are being tested on humans. This chapter describes the putative targets for TSE therapy and the current status of some of the therapeutic approaches under development for TSE treatment.

9.1 Targets for TSE therapy

Assuming that the hallmark event in the disease is the conversion of the α-helical PrP^C into the β-sheet-rich PrP^{Sc}, strategies that interfere with this conversion seem the most likely to lead to an effective therapy. At least seven different targets can be identified on the pathway from PrP^C to brain degeneration (Figure 9.1).

9.1.1 Reducing PrP expression

Using knockout animals, it has been shown that endogenous PrP^C is required for prion infection, although it is probably not necessary for any vital biological function [Bueler et al., 1993]. Thus it is conceivable that reducing expression of PrP^C could result in decrease of the disease. This can be achieved by altering the regulation of PrP gene expression or by using modern techniques of gene therapy, such as antisense oligonucleotide, RNAi, or genetically engineered ribozymes.

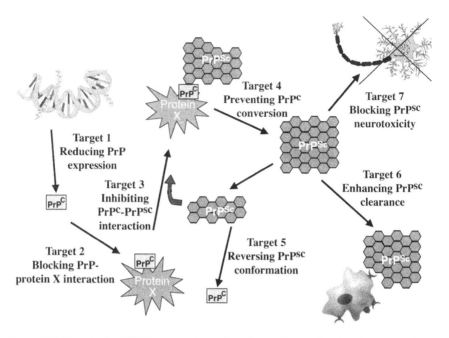

Figure 9.1 Targets for TSE therapy. Assuming that prion replication is a central process in the disease, several targets can be identified to block this process or its negative effect in the brain.

9.1.2   Blocking PrP–protein X interaction

As described in Chapter 3, experiments with transgenic animals and on biochemical studies of prion conversion *in vitro* suggest the likely existence of a cellular cofactor, known as protein X, that catalyzes the conversion process. The identity of this factor is not known. However, its site of interaction on the PrP^C molecule has been mapped [Kaneko et al., 1997], and inhibitors have already been designed to prevent the binding between PrP^C and protein X [Perrier et al., 2000]. Although this is an interesting target for therapy, it would be difficult to produce active molecules until the nature of the factor and its function are elucidated.

9.1.3   Inhibiting the interaction between PrP^C and PrP^{Sc}

In cell-free conversion studies, it has been shown that the first step in the conversion process is the specific interaction between the two forms of PrP [Horiuchi et al., 2000]. Therefore, blocking the binding of PrP^C to PrP^{Sc} should result in preventing prion replication. The internal sequence 106–141 of PrP has been associated with the interaction between PrP^C and PrP^{Sc}, and peptides spanning this sequence are capable of preventing PrP conversion *in vitro* [Chabry et al., 1998].

9.1.4   Preventing PrP^C conversion

Prevention of $\alpha \rightarrow \beta$ conformational change of PrP^C is the approach directed to what could be the first pathological process in TSE. Compounds able to overstabilize the structure of PrP^C may be effective inhibitors of prion replication. Stabilization of the native folding of PrP^C has been reported in prion-infected neuroblastoma cells treated with chemical chaperones (reagents known to stabilize native conformation of proteins), resulting in prevention of PrP misfolding [Tatzelt et al., 1996]. Protein engineering has been proposed as an approach to create sequence-modified proteins with higher stability, lower tendency to misfold, and the ability to trans-suppress the aggregation of wild-type protein [Villegas et al., 2000]. Indeed, several mutations or deletions in PrP have been shown to impair prion replication [Holscher et al., 1998; Perrier et al., 2002].

9.1.5   Reversing PrP^{Sc} conformation

Assuming the existence of two alternative stable conformations of PrP, it should be possible to find compounds able to destabilize the β-sheet structure of PrP^{Sc}, promoting the conversion of the protein back into the normal form. This approach is particularly exciting because it may produce benefit regardless of whether neurodegeneration is caused by a loss of function of PrP^C or by a gain of a toxic activity by PrP^{Sc}. In addition, since no early diagnosis is yet available, it has the potential to correct the preexisting misfolding of the protein.

9.1.6   Enhancing PrP^{Sc} clearance

It has been suggested that PrP^{Sc} accumulation is dependent on an imbalance between formation and clearance of the abnormally folded protein and that the disease might be due to defects in the proteasomal degradation of proteins [Dimcheff et al., 2003]. Perhaps the most promising strategy to increase the clearance of misfolded proteins is the immunization approach, first described to remove amyloid deposits in Alzheimer's disease [Schenk et al., 1999]. Recent studies have shown that active or passive immunization as well as transgenic expression of anti-PrP antibodies leads to a reduction of prion replication *in vitro* and *in vivo* [Enari et al., 2001; Heppner et al., 2001; Peretz et al., 2001; Sigurdsson et al., 2002].

9.1.7   Blocking PrP^{Sc} toxicity

Assuming that neurodegeneration in TSE is caused by a gain of toxic activity of the misfolded protein, a relevant target for therapy would be to block the mechanism by which PrP^{Sc} induces cell damage. As described in Chapter 7, PrP^{Sc} likely induces neuronal death by producing endoplasmic reticulum (ER) stress and activation of the caspase-12 pathway. This apoptosis pathway has its own endogenous defense mechanism consisting of the up-regulation of chaperone proteins. Our recent studies demonstrate that overexpression of the ER chaperone Grp58 leads to prevention of neuronal death triggered by PrP^{Sc} [Hetz et al., 2003]. Therefore, strategies aimed to boost this protective response may lead to a delay of brain degeneration in TSE. Alternatively, blocking caspase-12 activation may also result in a decrease of neuronal death. Interestingly, caspase-12 is one of the few members of the caspase family that does not seem to be involved in normal cellular homeostasis, but rather is associated only with pathological conditions [Mehmet, 2000]. Indeed, caspase-12-null animals are completely viable but are less suscepti-ble to neurodegeneration induced by misfolded proteins [Nakagawa et al., 2000]. However, as described in Chapter 7, neuronal death is not the only pathological alteration in the brain of TSE individuals, and longitudinal studies in rodent models have shown that early clinical signs of the disease occur before detectable neuronal loss.

9.2   Compounds under development for TSE treatment

Diverse compounds have been reported to alter prion replication *in vitro* or *in vivo* (Figure 9.2). Following is a brief description of some of these com-pounds and the current status of their development.

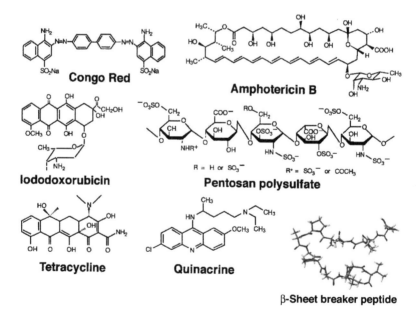

Figure 9.2 Structure of some chemical molecules that have been shown active in altering prion replication *in vitro* or *in vivo*.

9.2.1 Congo red, anthracyclines, and tetracycline

Several drugs are capable of interaction with PrP^{Sc} and intercalation into β-sheet structures, thus destabilizing the pathological conformation of PrP [Dormont, 2003; Rossi et al., 2003; Weissmann and Aguzzi, 2005]. Congo red is perhaps the prototype of a β-sheet intercalator and amyloid inhibitor. This compound binds specifically to aggregated β-sheet structures, regardless of the protein sequence, and has been shown to be active in blocking protein aggregation in a variety of systems [Klunk et al., 1989]. In TSEs, Congo red inhibits PrP^{Sc} accumulation in chronically infected cells [Caughey et al., 1993]. However, its *in vivo* effects remain controversial, with data showing a clear delay of disease onset in some experiments but no significant effect in others [Ingrosso et al., 1995; Milhavet et al., 2000; Poli et al., 2004]. In addition, Congo red is highly toxic and has a low permeability across the blood–brain barrier. For these reasons, attempts have been made to develop Congo red derivatives with enhanced potency and bioavailability and reduced toxicity [Klunk et al., 1998; Poli et al., 2003].

Iododoxorubicin is an anthracyclin that induces an increase in survival of hamsters when coinoculated with the 263 K scrapie agent [Tagliavini et al., 1997]. This compound was previously shown to effectively arrest systemic deposition of amyloid aggregates [Merlini et al., 1995]. The mechanism of action of this compound is not known, and in addition the drug cannot cross the blood–brain barrier. Therefore, its use in daily practice is not possible.

Because of its structural homology with the aglycone moiety of iodo-doxorubicin and its ability to cross the blood–brain barrier, tetracycline has been tested in rodent models [Forloni et al., 2001]. When coincubated with the inoculum, there was a significant delay in the onset of PrP^{Sc} accumulation inside the central nervous system (CNS) and the appearance of the clinical symptoms [Forloni et al., 2002]. The safety of this compound, proven through many years of use as an antibiotic as well as its good pharmacological properties, make it an interesting candidate for further development.

9.2.2 Polyanions

Polyanions are highly charged molecules known to inhibit cellular entry of several viruses by a nonspecific indirect mechanism involving interaction with the plasma membrane [Dormont, 2003]. Diverse polyanions, including dextran sulfate, pentosan sulfate, suramine, and HPA23 have shown an ability to delay the onset of scrapie in experimental rodent models [Dealler and Rainov, 2003; Kimberlin and Walker, 1986]. In general, efficiency requires early administration of polyanions either before or soon after peripheral infection and depends upon the degree of sulfatation of the drug. However, a recent report showed that when infused intraventricularly, pentosan polysulfate at high levels extended the survival of mice and decreased PrP^{Sc} deposition even when administered late after infection [Doh-Ura et al., 2004]. This is one of the few molecules that has shown such activity, and it is thus highly promising for the treatment of prion disease. Currently, a small clinical trial in humans affected by variant Creutzfeldt-Jakob disease (vCJD) is ongoing [Dyer, 2003].

9.2.3   Polyene antibiotics

Polyene antibiotics, such as amphotericin B, are antifungal agents thought to interact with ergosterol in the fungal envelope and with cholesterol present in mammalian cell membranes [Hartsel and Bolard, 1996]. Amphotericin B and its derivatives are some of the molecules that have shown the most significant effects on delaying prion diseases in animal models [Adjou et al., 1997]. The mechanism of action seems to be related to an ability to alter the integrity of the cholesterol-rich lipid-raft fraction of the cell membrane. It is known that both PrP^C and PrP^{Sc} are located in lipid-rafts domains, and there is evidence that the conversion process occurs in this membrane fraction. Therefore, any drug that could interfere with raft biology (modification of chemical composition, physical alteration) could theoretically impair prion replication. Amphotericin B and one of its less toxic derivatives, MS 8209, have been evaluated as anti-TSE agents in several host-strain combinations in infected rodents. A significant increase in survival time was observed in all experiments, particularly for MS 8209, with up to a 100% increase in survival of infected hamsters [Adjou et al., 1995; Adjou et al., 1999; Demaimay et al., 1997; Demaimay et al., 1999]. The therapeutic effect could be correlated with a delay in PrP^{Sc} accumulation in the brain. However, an earlier attempt to treat a CJD patient with amphotericin B was unsuccessful [Masullo et al., 1992].

Another polyene antibiotic, filipin, has also been shown to decrease PrP^{Sc} accumulation significantly in chronically infected cells [Marella et al., 2002]; this effect was associated with a reduction in PrP endocytosis.

9.2.4   Chlorpromazine and quinacrine

A strategy used by Prusiner's group to identify candidate drugs for TSE treatment was to screen molecules that have been approved for treating human beings affected by other disease indications. In this way, molecules that have good pharmacological and safety characteristics and that can cross the blood–brain barrier were selected for a screening study using chronically infected cells. Using this approach, Korth and colleagues showed that chlorpromazine and quinacrine can cure infected cells *in vitro* [Korth et al., 2001]. However, subsequent animal experiments failed to demonstrate efficacy in the treatment of TSEs [Barret et al., 2003; Collins et al., 2002], even after intraventricular infusion [Doh-Ura et al., 2004]. In addition, no clear therapeutic benefit was seen following quinacrine treatment of 32 patients affected by CJD [Haik et al., 2004], although some transient improvement occasionally occurred [Nakajima et al., 2004].

9.2.5 β-Sheet-breaker peptides

It has been proposed that short synthetic peptides containing the self-recognition motif of PrP and engineered to destabilize the abnormal conformation might be useful for TSE therapy [Soto, 1999; Soto et al., 2000]. These peptides, called β-sheet breakers, are designed to be similar to the PrP sequence responsible for self-association and contain residues that would not fit in the β-sheet structure of PrP^{Sc}. Binding of these peptides to PrP^C may prevent its acquisition of β-sheet conformation, and their interaction with PrP^{Sc} may induce the destabilization of the abnormal PrP conformation. Therefore, these compounds are designed to both inhibit PrP conversion and to correct the misfolding of previously converted protein. The β-sheet-breaker peptides were created maintaining a partial homology to the PrP fragment 114–122 [Soto et al., 2000], since several pieces of evidence suggest that this region plays a central role in PrP self-recognition and in the conversion PrP^C → PrP^{Sc}. To disrupt β-sheet formation, proline residues were added, because the occurrence of this residue in a β-pleated structure is energetically unfavorable due to the constraints on its ability to support the required peptide backbone conformation [Wood et al., 1995].

The ability of β-sheet-breaker peptides to convert protease-resistant human and animal PrP^{Sc} into a protease-sensitive PrP^C-like form was tested using PrP^{Sc} purified from mice affected by scrapie as well as from sporadic and new variant CJD brains. The results of these studies demonstrated for the first time the reversion of the biochemical and structural properties of PrP^{Sc} to a state similar to PrP^C by treatment with a 13-residue β-sheet-breaker peptide [Soto et al., 2000]. Similar results have been obtained recently with shorter derivatives and with peptides containing modifications aimed to enhance pharmacological properties [Soto, 2003]. *In vitro* activity has also been studied using an *in vitro* conversion system based on cyclic amplification of prion protein misfolding and in chronically infected neuroblastoma cells.

In vivo studies using experimental scrapie models in mice demonstrated that *ex vivo* treatment of prion infectious material with β-sheet-breaker peptides resulted in a delay in the appearance of the clinical symptoms and a 90 to 95% reduction in infectivity [Soto et al., 2000]. The reduction in infectivity correlated with the decrease in the β-sheet content and the protease resistance of PrP^{Sc} induced by incubation with the peptide *in vitro*. Our findings suggest that β-sheet-breaker peptides or derivatives thereof could represent a new therapeutic approach to prevent and reverse the PrP conformational changes associated with the pathogenesis of TSE.

The major drawback to this approach is that the peptide nature of the compounds makes them easily degraded, with low permeability across biological barriers. Some of these weaknesses can be minimized by introducing specific chemical modifications aimed to protect them from peptidase degradation and increase membrane permeability [Adessi and Soto, 2002]. Alternatively, a peptide mimetic could be designed maintaining the structure and activity characteristics of the lead peptide in a nonpeptide chemical molecule.

9.3 Immunization approach

One of the intriguing characteristics of TSE is the lack of detectable specific immune response in infected individuals. This is believed to be related to the particular nature of the infectious agent, which is composed of a misfolded version of a host-encoded protein. Therefore, immune intervention has often been considered of little use in TSE. However, several groups have recently been attempting to develop a vaccine for TSE, inspired by the pioneering work of Schenk and colleagues, who demonstrated that immunization of transgenic mice models of Alzheimer's disease induces a significant decrease of protein aggregation and neurodegeneration [Schenk et al., 1999].

Antibodies directed against some PrP epitopes are able to cure chronically infected cells [Enari et al., 2001; Peretz et al., 2001]. *In vivo*, transgenic mice harboring a PrP antibody μ-chain are significantly less susceptible to peripheral infection with the scrapie agent [Heppner et al., 2001]. Anti-PrP antibodies were found to inhibit formation of protease-resistant PrP in a cell-free system [Horiuchi and Caughey, 1999]. Moreover, vaccination with recombinant PrP prior to or just after infection led to a significant delay of the disease onset [Sigurdsson et al., 2002]. Finally, passive immunization with antibodies directed against PrP (epitopes 91–110 and 149–159) increased animal survival [Sigurdsson et al., 2003]. However, passive immunization failed to confer protection if treatment was started after the onset of clinical symptoms, so it might be a better candidate for prophylaxis than for therapy of TSEs. Active immunization may be more effective, as with most antiviral vaccines, but it is rendered exceedingly difficult by the tolerance to PrP^C [Polymenidou et al., 2004].

Despite the promising results obtained in animals, caution has to be taken in using this approach in humans. In fact, the immunization approach for Alzheimer's therapy was abruptly stopped when several patients developed complicated inflammatory side effects in a phase II clinical trial [Orgogozo et al., 2003]. Indeed, a recent report showed that intracerebral injection of anti-PrP antibodies specific to certain epitopes at high concentrations provoked degeneration of hippocampal and cerebellar neurons [Solforosi et al., 2004].

9.4 Concluding remarks

Unfortunately, little progress has been made in finding new efficient therapies for these devastating diseases. Although several compounds have been shown to delay the onset of the disease in animal models, none of them is actively in the clinical stage. The rapid progression of the disease once the clinical symptoms appear and the large extent of brain damage already made at this time impose serious obstacles for a therapy. In addition, ethical considerations have to be balanced when attempting to stabilize the disease and prolong the life of people at the clinical stage of the illness. It seems that the best option will be to develop diagnostic procedures that enable the treatment to begin at a time before any severe brain damage has occurred. Until this is achieved, the likelihood of developing an efficient therapy for TSEs seems low.

References* **

Adessi, C. and Soto, C., Converting a peptide into a drug: strategies to improve stability and bioavailability, *Curr. Med. Chem.*, 9, 963–978, 2002.

Adjou, K.T. et al., MS-8209, a new amphotericin B derivative, provides enhanced efficacy in delaying hamster scrapie, *Antimicrob. Agents Chemother.*, 39, 2810–2812, 1995.

Adjou, K.T. et al., Probing the dynamics of prion diseases with amphotericin B, *Trends Microbiol.*, 5, 27–31, 1997.

Adjou, K.T. et al., MS-8209, a water-soluble amphotericin B derivative, affects both scrapie agent replication and PrPres accumulation in Syrian hamster scrapie, *J. Gen. Virol.*, 80, 1079–1085, 1999.

Barret, A. et al., Evaluation of quinacrine treatment for prion diseases, *J. Virol.*, 77, 8462–8469, 2003.

Bueler, H. et al., Mice devoid of PrP are resistant to scrapie, *Cell*, 73, 1339–1347, 1993.

*Caughey, B., Ernst, D., and Race, R.E., Congo red inhibition of scrapie agent replication, *J. Virol.*, 67, 6270–6272, 1993. (One of the first reports of a molecule inhibiting prion replication *in vitro*.)

Chabry, J., Caughey, B., and Chesebro, B., Specific inhibition of *in vitro* formation of protease-resistant prion protein by synthetic peptides, *J. Biol. Chem.*, 273, 13203–13207, 1998.

Collins, S.J. et al., Quinacrine does not prolong survival in a murine Creutzfeldt-Jakob disease model, *Ann. Neurol.*, 52, 503–506, 2002.

Dealler, S. and Rainov, N.G., Pentosan polysulfate as a prophylactic and therapeutic agent against prion disease, *IDrugs*, 6, 470–478, 2003.

*Demaimay, R. et al., Late treatment with polyene antibiotics can prolong the survival time of scrapie-infected animals, *J. Virol.*, 71, 9685–9689, 1997. (Describes one of the few experiments in which treatment at late stage produces a significant delay of disease onset.)

Demaimay, R., Race, R., and Chesebro, B., Effectiveness of polyene antibiotics in treatment of transmissible spongiform encephalopathy in transgenic mice expressing Syrian hamster PrP only in neurons, *J. Virol.*, 73, 3511–3513, 1999.

Dimcheff, D.E., Portis, J.L., and Caughey, B., Prion proteins meet protein quality control, *Trends Cell Biol.*, 13, 337–340, 2003.

*Doh-Ura, K. et al., Treatment of transmissible spongiform encephalopathy by intraventricular drug infusion in animal models, *J. Virol.*, 78, 4999–5006, 2004. (This is an interesting study that evaluates the therapeutic efficacy of several drug candidates for TSE. The studies are done using experimental scrapie in rodents at late stages of the disease.)

**Dormont, D., Approaches to prophylaxis and therapy, *Br. Med. Bull.*, 66, 281–292, 2003.

* Highlights primary articles of outstanding importance and quality, including a short description of the findings.
** Highlights comprehensive review articles similar to the topic of this chapter.

Dyer, C., Second vCJD patient to receive experimental treatment, *BMJ*, 327, 886, 2003.

Enari, M., Flechsig, E., and Weissmann, C., Scrapie prion protein accumulation by scrapie-infected neuroblastoma cells abrogated by exposure to a prion protein antibody, *Proc. Natl. Acad. Sci. USA*, 98, 9295–9299, 2001.

Forloni, G. et al., Anti-amyloidogenic activity of tetracyclines: studies *in vitro*, *FEBS Lett.*, 487, 404–407, 2001.

Forloni, G. et al., Tetracyclines affect prion infectivity, *Proc. Natl. Acad. Sci. USA*, 99, 10849–10854, 2002.

Haik, S. et al., Compassionate use of quinacrine in Creutzfeldt-Jakob disease fails to show significant effects, *Neurology*, 63, 2413–2415, 2004.

Hartsel, S. and Bolard, J., Amphotericin B: new life for an old drug, *Trends Pharmacol. Sci.*, 17, 445–449, 1996.

Heppner, F.L. et al., Prevention of scrapie pathogenesis by transgenic expression of anti-prion protein antibodies, *Science*, 294, 178–182, 2001.

Hetz, C. et al., Caspase-12 and endoplasmic reticulum stress mediate neurotoxicity of pathological prion protein, *EMBO J.*, 22, 5435–5445, 2003.

Holscher, C., Delius, H., and Burkle, A., Overexpression of nonconvertible PrP^C delta114-121 in scrapie-infected mouse neuroblastoma cells leads to trans-dominant inhibition of wild-type PrP(Sc) accumulation, *J. Virol.*, 72, 1153–1159, 1998.

Horiuchi, M. and Caughey, B., Specific binding of normal prion protein to the scrapie form via a localized domain initiates its conversion to the protease-resistant state, *EMBO J.*, 18, 3193–3203, 1999.

Horiuchi, M. et al., Interactions between heterologous forms of prion protein: binding, inhibition of conversion, and species barriers, *Proc. Natl. Acad. Sci. USA*, 97, 5836–5841, 2000.

Ingrosso, L., Ladogana, A., and Pocchiari, M., Congo red prolongs the incubation period in scrapie-infected hamsters, *J. Virol.*, 69, 506–508, 1995.

Kaneko, K. et al., Evidence for protein X binding to a discontinuous epitope on the cellular prion protein during scrapie prion propagation, *Proc. Natl. Acad. Sci. USA*, 94, 10069–10074, 1997.

Kimberlin, R.H. and Walker, C.A., Suppression of scrapie infection in mice by heteropolyanion 23, dextran sulfate, and some other polyanions, *Antimicrob. Agents Chemother.*, 30, 409–413, 1986.

Klunk, W.E., Pettegrew, J.W., and Abraham, D.J., Quantitative evaluation of Congo red binding to amyloid-like proteins with a beta-pleated sheet conformation, *J. Histochem. Cytochem.*, 37, 1273–1281, 1989.

Klunk, W.E. et al., Chrysamine-G, a lipophilic analogue of Congo red, inhibits a beta-induced toxicity in PC12 cells, *Life Sci.*, 63, 1807–1814, 1998.

Korth, C. et al., Acridine and phenothiazine derivatives as pharmacotherapeutics for prion disease, *Proc. Natl. Acad. Sci. USA*, 98, 9836–9841, 2001.

Marella, M. et al., Filipin prevents pathological prion protein accumulation by reducing endocytosis and inducing cellular PrP release, *J. Biol. Chem.*, 277, 25457–25464, 2002.

Masullo, C. et al., Failure to ameliorate Creutzfeldt-Jakob disease with amphotericin B therapy, *J. Infect. Dis.*, 165, 784–785, 1992.

Mehmet, H., Caspases find a new place to hide, *Nature*, 403, 29–30, 2000.

Merlini, G. et al., Interaction of the anthracycline 4'-iodo-4'-deoxydoxorubicin with amyloid fibrils: inhibition of amyloidogenesis, *Proc. Natl. Acad. Sci. USA*, 92, 2959–2963, 1995.

Milhavet, O. et al., Effect of Congo red on wild-type and mutated prion proteins in cultured cells, *J. Neurochem.*, 74, 222–230, 2000.

Nakagawa, T. et al., Caspase-12 mediates endoplasmic-reticulum-specific apoptosis and cytotoxicity by amyloid-beta, *Nature*, 403, 98–103, 2000.

Nakajima, M. et al., Results of quinacrine administration to patients with Creutzfeldt-Jakob disease, *Dement. Geriatr. Cogn Disord.*, 17, 158–163, 2004.

Orgogozo, J.M. et al., Subacute meningoencephalitis in a subset of patients with AD after Abeta42 immunization, *Neurology*, 61, 46–54, 2003.

*Peretz, D. et al., Antibodies inhibit prion propagation and clear cell cultures of prion infectivity, *Nature*, 412, 739–743, 2001. (One of the earliest reports indicating that immunotherapeutic approaches may be promising for the treatment of prion diseases.)

Perrier, V. et al., Dominant-negative inhibition of prion replication in transgenic mice, *Proc. Natl. Acad. Sci. USA*, 99, 13079–13084, 2002.

Perrier, V. et al., Mimicking dominant negative inhibition of prion replication through structure-based drug design, *Proc. Natl. Acad. Sci. USA*, 97, 6073–6078, 2000.

Poli, G. et al., *In vitro* evaluation of the anti-prionic activity of newly synthesized Congo red derivatives, *Arzneimittelforschung*, 53, 875–888, 2003.

Poli, G. et al., Evaluation of anti-prion activity of Congo red and its derivatives in experimentally infected hamsters, *Arzneimittelforschung*, 54, 406–415, 2004.

Polymenidou, M. et al., Humoral immune response to native eukaryotic prion protein correlates with anti-prion protection, *Proc. Natl. Acad. Sci. USA*, 101, 14670–14676, 2004.

**Rossi, G. et al., Therapeutic approaches to prion diseases, *Clin. Lab. Med.*, 23, 187–208, 2003.

Schenk, D. et al., Immunization with amyloid-beta attenuates Alzheimer-disease-like pathology in the PDAPP mouse, *Nature*, 400, 173–177, 1999.

Sigurdsson, E.M. et al., Immunization delays the onset of prion disease in mice, *Am. J. Pathol.*, 161, 13–17, 2002.

Sigurdsson, E.M. et al., Anti-prion antibodies for prophylaxis following prion exposure in mice, *Neurosci. Lett.*, 336, 185–187, 2003.

Solforosi, L. et al., Cross-linking cellular prion protein triggers neuronal apoptosis *in vivo*, *Science*, 303, 1514–1516, 2004.

Soto, C., Alzheimer's and prion disease as disorders of protein conformation: implications for the design of novel therapeutic approaches, *J. Mol. Med.*, 77, 412–418, 1999.

Soto, C., β-Sheet breaker peptides for the treatment of transmissible spongiform encephalopathies, in *New Perspectives on Prion Therapeutics*, Lehmann, S., Ed., Editions de Conde, Paris, 2003, pp. 145–154.

*Soto, C. et al., Reversion of prion protein conformational changes by synthetic beta-sheet breaker peptides, *Lancet*, 355, 192–197, 2000. (Describes the application of the β-sheet-breaker approach to the treatment of TSE.)

Tagliavini, F. et al., Effectiveness of anthracycline against experimental prion disease in Syrian hamsters, *Science*, 276, 1119–1122, 1997.

Tatzelt, J., Prusiner, S.B., and Welch, W.J., Chemical chaperones interfere with the formation of scrapie prion protein, *EMBO J.*, 15, 6363–6373, 1996.

Villegas, V. et al., Protein engineering as a strategy to avoid formation of amyloid fibrils, *Protein Sci.*, 9, 1700–1708, 2000.

**Weissmann, C. and Aguzzi, A., Approaches to therapy of prion diseases, *Annu. Rev. Med.*, 56, 321–344, 2005.

Wood, S.J. et al., Prolines and amyloidogenicity in fragments of the Alzheimer's peptide beta/A4, *Biochemistry,* 34, 724–730, 1995.

chapter ten

Cyclic amplification of prion protein misfolding: rationale, applications, and perspectives

Prion replication is a slow process *in vivo*. We have recently described a procedure to induce accelerated prion replication *in vitro* based on the cyclic amplification of prion protein misfolding (PMCA) [Saborio et al., 2001]. This procedure, conceptually analogous to DNA amplification by PCR, has tremendous implications for research and diagnosis [Soto, 2002; Soto et al., 2002]. This chapter describes the principles behind this novel technology, the current status of development, and its applications for understanding the molecular basis of the prion infectious agent (PrP^{Sc}) and for developing ultrasensitive diagnostic procedures.

10.1 The rationale behind PMCA

The prion concept by which a protein can replicate in the absence of nucleic acid by transmitting its folding and biochemical features to other molecules of the protein is a revolutionary concept in biology, since until recently it was believed that only nucleic acids were capable of replication and transmission of biological information [Soto and Saborio, 2001]. The self-replication capability of DNA has been used to amplify *in vitro* pieces of DNA by the polymerase chain reaction (PCR). This amplification process allows synthesis of large amounts of a specific DNA starting with minute amounts (frequently undetectable) of template. The question that has been in the mind of many biochemists is how to develop a system conceptually analogous to PCR to amplify the properties and activity of proteins. The recognition of proteins (such as prion proteins) with the capability to propagate their biological properties opens a new possibility to develop such a system.

We have recently reported a new technology named protein misfolding cyclic amplification (PMCA) that can multiply minute quantities of PrP^{Sc} at expenses of large amounts of PrP^C [Saborio et al., 2001]. This technology mimics the prion replication process at an accelerated pace [Soto et al., 2002]. As discussed in Chapter 3, although the detailed mechanism of prion replication is not totally understood, it seems to involve a close interaction between both protein isoforms facilitated by an as-yet-unidentified conversion factor. PrP^{Sc} has been described as an oligomer of variable size that acts as a seed to recruit molecules of partially misfolded PrP^C, stabilizing the misfolding by incorporation into the oligomer [Caughey et al., 1997; Jarrett and Lansbury, Jr., 1993] (Figure 10.1). Thus, the PrP^{Sc} oligomer is elongated at the ends as new molecules of PrP^C are converted and incorporated. The kinetics of such nucleated conformational conversion is limited by the number of seeds present in the sample [Jarrett and Lansbury, Jr., 1993; Masel et al., 1999; Soto et al., 2002]. This rate-limiting process might explain in part the long period of time needed *in vivo* to generate a concentration of PrP^{Sc} high enough to trigger neurodegeneration.

The idea behind PMCA is to combine the growing of PrP^{Sc} polymers with a step in which the number of conversion units (seeds) are multiplied (Figure 10.1). In a cyclic manner, a minute quantity of PrP^{Sc} is incubated with

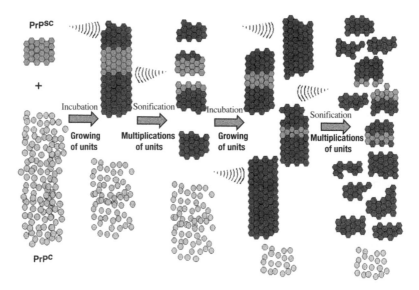

Figure 10.1 (See color insert after p. 114.) Diagrammatic representation of PMCA. Cyclic amplification consists of subjecting a sample containing minute quantities of PrP^{Sc} and a large excess of PrP^C to cycles of incubation/sonication. During incubation, PrP^{Sc} aggregates grow by converting and incorporating PrP^C into the polymer. Sonication is used to break down large aggregates into many smaller pieces to multiply the number of converting units. PMCA cycles can be repeated as many times as needed to amplify undetectable quantities of PrP^{Sc} to a level of easy detection.

excess PrP^C to enlarge the PrP^{Sc} aggregates, which are then sonicated to generate multiple smaller units for the continued formation of new PrP^{Sc} [Saborio et al., 2001; Soto, 2002; Soto et al., 2002]. We have reported proof-of-concept experiments in which the sensitivity of PrP^{Sc} detection was increased 10- to 60-fold [Saborio et al., 2001], and the technology was recently applied to replicate the misfolded protein from diverse species [Soto et al., 2005]. A modest level of amplification has also been observed without son-ication [Deleault et al., 2003; Lucassen et al., 2003; Saborio et al., 2001], and the extent of conversion depends upon the number of PMCA cycles [Bie-schke et al., 2004; Piening et al., 2005; Saborio et al., 2001].

PMCA requires a PrP^C substrate for conversion and additional cellular factors essential for PrP replication. Because the brain is the target organ for TSE and the tissue where most of the conversion takes place, we used healthy brain homogenate as a source of PrP^C substrate and conversion factors. The conditions of the solution were selected to be close to the physiological environment in terms of pH and ionic strength in an effort to avoid affecting the interactions among the different components. PrP^C present in the sample does not acquire protease resistance by PMCA if a template of PrP^{Sc} is not present [Saborio et al., 2001]. In the same way, the PrP^{Sc} initial signal is not modified with the cycles of incubation-sonication in the absence of an ade-quate substrate (PrP^C) [Saborio et al., 2001].

10.2 Applications of PMCA in prion diagnosis

As described in Chapter 8, the development of tests that can effectively identify animals and people incubating the different forms of TSE is a top priority [Brown et al., 2001; Schiermeier, 2001; Soto, 2004]. PrP^{Sc} is the only validated surrogate marker for the disease [Brown et al., 2001; Prusiner, 1998], but the problem for diagnosis is that PrP^{Sc} is abundant only in the brain at late stages of the disease. However, several lines of evidence indicate that prions are also present in minute amounts in peripheral tissues, such as lymphoid organs and blood [Aguzzi, 2000; Brown et al., 2001; Wadsworth et al., 2001]. PMCA offers the possibility of amplifying the amount of PrP^{Sc} in a sample, making its detection easier by existing methods.

We recently reported that PMCA allowed detection of PrP^{Sc} in the brain of presymptomatic hamsters, enabling a clear identification of infected ani-mals as early as 2 weeks after inoculation [Soto et al., 2005]. More impor-tantly, PMCA demonstrated the presence of PrP^{Sc} in an experimentally infected cow 32 months postinoculation that showed no clinical signs and was negative by standard Western blot analysis [Soto et al., 2005]. We also showed that PMCA can be applied successfully to a variety of brain samples from experimental and natural TSEs of humans and animals, including 263K hamster scrapie, RML mouse scrapie, sheep scrapie, goat scrapie, bovine spongiform encephalopathy (BSE), and sporadic and variant Creutzfeldt-Jakob disease (CJD) [Soto et al., 2005]. Thus, the same principle operates with a number of prion strains, and all PrP^{Sc} isoforms seem to be capable of converting a large excess of PrP^C *in vitro* under specific conditions. Moreover, the Western blot profile and the protease sensitivity of newly formed PrP^{Sc} are identical to those of PrP^{Sc} used for the reaction. The mechanism by which PrP^{Sc} can faithfully transfer its biochemical properties to PrP^C remains to be elucidated.

The application of the PMCA technology for large-scale early biochemical diagnosis in biological fluids (e.g., blood) is dependent upon designing an automated, high-throughput system that will enable an increase of sensitivity of three to five orders of magnitude over existing technologies [Soto, 2004]. With this aim, we have developed an automated PMCA system in which the sonication is carried out in a microplate horn block sonicator that can be programmed for automatic operation [Castilla et al., 2004; Castilla et al., 2005; Saa et al., 2004]. This improvement not only decreases processing time and allows routine processing of many more samples than a single-probe sonicator, but also prevents loss of material. Cross-contamination is eliminated, since there is no direct probe intrusion into the sample. The automated and optimized PMCA procedure enables a very sensitive detection. We recently reported that 140 PMCA cycles leads to a 6600-fold increase of sensitivity over standard detection methods [Castilla et al., 2005b]. Strikingly, two successive rounds of 100 PMCA cycles resulted in a 10-million-fold increase in sensitivity and a capability to detect as little as 8,000 molecules of PrP^{Sc} [Castilla et al., 2005b]. This sensitivity is even higher than the most efficient infectious bioassay in animals. One round of 96 PMCA cycles detected PrP^{Sc} in the spleen of infected animals with 100% sensitivity and specificity. Six rounds of PMCA cycling enabled detection of PrP^{Sc} in 89% of samples of blood from hamsters infected with scrapie prions [Castilla et al., 2005b]. These findings suggest that automated serial PMCA enables efficient, specific, and rapid detection of prions in a variety of samples, offering great promise for developing a noninvasive early diagnosis of prion diseases.

10.3 In vitro *generation of infectious prions by PMCA*

A crucial issue in TSE is the unprecedented nature of the infectious agent, which according to the prion hypothesis is composed of a single protein that propagates in the absence of nucleic acid [Prusiner, 1998]. Although strong evidence in favor of this hypothesis has been reported in the last few years, it is still a matter of controversy among many scientists [Chesebro, 1998; Mestel, 1996; Soto and Castilla, 2004]. Perhaps the most important missing evidence for the protein-only hypothesis is the generation of infectivity in the test tube [Soto and Castilla, 2004]. PMCA provides a unique opportunity to evaluate the infectious properties of PrP^{Sc} generated *in vitro* because, after amplification under optimal conditions, >99% of protease-resistant protein is composed of newly produced PrP^{Sc}. The high yield following conversion is essential to distinguish newly generated infectivity from that used to initiate the reaction.

The PMCA procedure has been further optimized to serially amplify PrP^{Sc} *in vitro* indefinitely. After many rounds of PMCA following serial dilution of brain infectious material, we have been able to generate large quantities of newly synthesized PrP^{Sc} in the absence of any molecules of the brain-derived misfolded protein [Castilla et al., 2005a]. These results confirm a central facet of the prion hypothesis, which is that prion replication is a cyclical process and that newly produced PrP^{Sc} can further propagate the protein misfolding. In this way, prions can maintain replication across animals and generations. Our results show that this process of autocatalytic generation of PrP^{Sc} can be mimicked *in vitro* by PMCA, indicating that prions can be "cultured" *in vitro* indefinitely, thus enabling the detailed study of their properties. *In vitro*-produced PrP^{Sc} exhibited strikingly similar properties compared with the protein isolated from scrapie-affected animals, including secondary structure, Western blot profile, protease-resistance, detergent insolubility, resistance to denaturation by heat and chaotropic agents, and aggregation into rodlike structures [Castilla et al., 2005a].

The serial replication of PrP^{Sc} by PMCA provides a perfect system to evaluate the infectious properties of *in vitro*-generated PrP^{Sc}, because after many rounds of amplification following serial dilution of PrP^{Sc} inoculum, we are able to produce a preparation of misfolded protein that is biochemically and structurally identical to brain-derived PrP^{Sc} but lacks any molecule of the initial scrapie-infectious material [Castilla et al., 2005a]. To determine the infectious capability of *in vitro*-generated protease-resistant PrP (PrP^{res}), groups of wild-type Syrian hamsters were inoculated intracerebrally (i.c.) with samples of PrP^{Sc} extracted from the brain or produced entirely *in vitro* and not containing any molecule of brain infectious material. The results clearly demonstrated that PMCA-generated PrP^{Sc} was infectious to wild-type animals (Figure 10.2A). Moreover, the disease produced by this infection had identical clinical, histological, and biochemical characteristics than the disease induced by inoculation with brain infectious agent (Figure 10.2B). The infectivity generated *in vitro* was stable over time and can be serially passed to animals by inoculation of the brain material [Castilla et al., 2005a].

10.4   Application of PMCA to understand the prion replication process

Another widely debated issue in the TSE field is the molecular mechanism of species barrier and prion strains [Clarke et al., 2001; Kascsak et al., 1991]. The transmission of BSE to humans has led to great concern regarding interspecies infectivity and the identification of tissues having a quantity of prions high enough to transmit the disease [Hill et al., 2000; Hunter, 2003; Wadsworth et al., 2001]. The molecular aspects that underlie the species-barrier phenomenon are still not understood. It has been shown that the sequence homology between the infectious PrP^{Sc} and the host prion protein plays a crucial role in determining species barrier [Telling et al., 1996]. It is clear that few amino acid

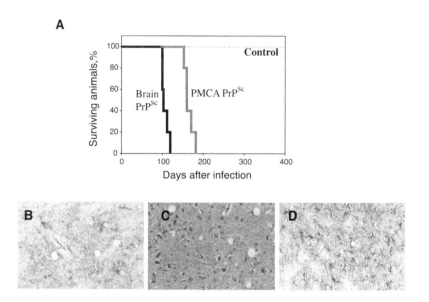

Figure 10.2 (See color insert after p. 114.) Infectious properties of *in vitro*-generated PrP^{Sc}: (A) Wild-type hamsters, inoculated with similar quantities of PrP^{Sc} either generated *in vitro* or derived from the brain, developed the disease. The graph shows the survival time for each animal in groups inoculated with each brain infectious material or with *in vitro*-generated PrP^{Sc} under conditions in which no molecules of inoculum PrP^{Sc} are present. (B) PrP^{Sc} accumulates in the brain of animals sick after inoculation with *in vitro*-generated infectivity. (C) Hematoxilin-eosin staining of the brain revealed vacuolation, which represents the typical spongiform degeneration observed in TSEs. (D) Astrocytes, another of the typical brain alterations observed in these animals, were studied by glial fibrillary acidic protein (GFAP) staining.

differences between both proteins can modify dramatically the incubation time and the course of the disease [Asante and Collinge, 2001; DeArmond and Prusiner, 1996]. So far, the investigation of the species barrier and the tissues carrying infectivity has been done using the biological assay of prion propagation in animals [Chen and Gambetti, 2002]. However, these studies are time consuming because it is necessary to wait for several months or even years until the animals develop the clinical symptoms. In addition, the assessment of the species barrier for transmission of prions to humans is compromised by the use of animal models. PMCA might provide a complement to the *in vivo* studies of the phenomenon of species barrier by combining PrP^{Sc} and PrP^{C} from different sources in distinct quantities and evaluating quantitatively the efficiency of conversion. In addition, the presence of minute quantities of prions in different tissues can be analyzed by attempting to convert brain PrP^{C} with a sample of a determined tissue from an infected individual. These studies should provide the basis for minimizing the risk of prion propagation. Our preliminary results indicate the possibility of using PMCA to cross the barrier between mice and hamsters to produce a misfolded protein that is infectious to the other species.

As described in Chapter 3, based initially on data with transgenic animals, it has been proposed that additional brain factors present in the host are essential for prion propagation [Telling et al., 1995]. We demonstrated previously that prion conversion does not occur under our experimental conditions when purified PrP^C and PrP^{Sc} are mixed and incubated [Saborio et al., 1999]. The conversion activity was recovered when the bulk of cellular proteins were added back to the sample [Saborio et al., 1999]. This finding provides direct evidence that other factors present in the brain are essential to catalyze prion propagation. PMCA may be useful as a biochemical assay to identify these conversion factors. Historically, the enzymatic activity of proteins has been used to facilitate their purification and characterization. Until now, there was no method available to isolate and study the cellular factors involved in the interaction or interconversion of PrP^C into PrP^{Sc}. Having the appropriate system to identify the elements required for prion conversion may offer new opportunities to study the molecular mechanism of PrP^{Sc} formation and the pathogenesis of TSE. Indeed, using PMCA we have already seen substantial progress on the identification of such factors. Following are some of the as-yet-unpublished observations regarding our efforts to identify this factor.

The conversion factor is most likely a protein, since conversion activity was not eliminated by treatment with nucleases, lipases, or enzymes that destroy carbohydrates. The conversion factor is preferentially expressed in the brain, although it is also present in lower concentrations in muscle, but not in other tissues. It is located in membrane lipid rafts, which are specialized regions of the membrane enriched in signaling proteins and where both PrP^C and PrP^{Sc} are also located [Russelakis-Carneiro et al., 2004; Vey et al., 1996].

10.5   Concluding remarks

PMCA represents the first opportunity to cultivate prions *in vitro* indefinitely and in a highly efficient way. This system should provide a means of dramatically advancing research into the molecular basis of prion replication and the nature of the infectious agent. This is important not only for the knowledge of the mechanism associated with this unprecedented infectious agent, but it also may lead to the development of novel strategies for therapy. In addition, PMCA provides a unique opportunity to increase spectacularly the sensitivity of PrP^{Sc} detection and thus may lead to development of ultrasensitive diagnosis of TSE. The future of PMCA seems bright, and there is hope that its principles can be applied to develop similar amplification procedures for other protein-misfolding processes implicated in other diseases.

References* **

Aguzzi, A., Prion diseases, blood and the immune system: concerns and reality, *Haematologica*, 85, 3–10, 2000.

Asante, E.A. and Collinge, J., Transgenic studies of the influence of the PrP structure on TSE diseases, *Adv. Protein Chem.*, 57, 273–311, 2001.

*Bieschke, J. et al., Autocatalytic self-propagation of misfolded prion protein, *Proc. Natl. Acad. Sci. USA*, 101, 12207–12211, 2004. (Application of the PMCA technology for reaching serial amplification of PrP^{Sc}. Unfortunately, due to the relatively low levels of amplification, the infectivity studies were inconclusive.)

Brown, P., Cervenakova, L., and Diringer, H., Blood infectivity and the prospects for a diagnostic screening test in Creutzfeldt-Jakob disease, *J. Lab Clin. Med.*, 137, 5–13, 2001.

**Castilla, J., Saa, P., and Soto, C., Cyclic amplification of prion protein misfolding, in *Methods and Tools in Bioscience and Medicine: Techniques in Prion Research*, Lehmann, S. and Grassi, J., Eds., Birkhauser Verlag, Basel, 2004, pp. 198–213.

*Castilla, J., Saa, P., Hetz, C., and Soto, C., *In vitro* generation of scrapie infectious prions, *Cell*, 121, 195–206, 2005a. (Demonstrates that *in vitro*-generated PrP^{Sc} is infectious. These findings represent some of the strongest evidence in favor of the prion hypothesis.)

Castilla, J., Saa, P., and Soto, C., Biochemical detection of prions in blood, *Nature Med.*, 11, 982–985, 2005b (Reports for the first time the detection of prions in blood).

Caughey, B. et al., Scrapie infectivity correlates with converting activity, protease resistance, and aggregation of scrapie-associated prion protein in guanidine denaturation studies, *J. Virol.*, 71, 4107–4110, 1997.

Chen, S.G. and Gambetti, P., A journey through the species barrier, *Neuron*, 34, 854–856, 2002.

Chesebro, B., BSE and prions: uncertainties about the agent, *Science*, 279, 42–43, 1998.

Clarke, A.R., Jackson, G.S., and Collinge, J., The molecular biology of prion propagation, *Philos. Trans. R. Soc. Lond B Biol. Sci.*, 356, 185–195, 2001.

DeArmond, S.J. and Prusiner, S.B., Transgenetics and neuropathology of prion diseases, *Curr. Top. Microbiol. Immunol.*, 207, 125–146, 1996.

*Deleault, N.R., Lucassen, R.W., and Supattapone, S., RNA molecules stimulate prion protein conversion, *Nature*, 425, 717–720, 2003. (An interesting report suggesting that the factor involved in the replication of PrP^{Sc} might be RNA.)

Hill, A.F. et al., Species-barrier-independent prion replication in apparently resistant species, *Proc. Natl. Acad. Sci. USA*, 97, 10248–10253, 2000.

Hunter, N., Scrapie and experimental BSE in sheep, *Br. Med. Bull.*, 66, 171–183, 2003.

Jarrett, J.T. and Lansbury, P.T., Jr., Seeding "one-dimensional crystallization" of amyloid: a pathogenic mechanism in Alzheimer's disease and scrapie? *Cell*, 73, 1055–1058, 1993.

Kascsak, R.J., Rubenstein, R., and Carp, R.I., Evidence for biological and structural diversity among scrapie strains, *Curr. Top. Microbiol. Immunol.*, 172, 139–152, 1991.

---

* Highlights primary articles of outstanding importance and quality, including a short description of the findings.
** Highlights comprehensive review articles similar to the topic of this chapter.

Lucassen, R., Nishina, K., and Supattapone, S., *In vitro* amplification of protease-resistant prion protein requires free sulfhydryl groups, *Biochemistry*, 42, 4127–4135, 2003.

Masel, J., Jansen, V.A., and Nowak, M.A., Quantifying the kinetic parameters of prion replication, *Biophys. Chem.*, 77, 139–152, 1999.

Mestel, R., Putting prions to the test, *Science*, 273, 184–189, 1996.

Piening, N. et al., Breakage of PrP aggregates is essential for efficient autocatalytic propagation of misfolded prion protein, *Biochem. Biophys. Res. Commun.*, 326, 339–343, 2005.

Prusiner, S.B., Prions, *Proc. Natl. Acad. Sci. USA*, 95, 13363–13383, 1998.

Russelakis-Carneiro, M. et al., Prion replication alters the distribution of synaptophysin and caveolin 1 in neuronal lipid rafts, *Am. J. Pathol.*, 165, 1839–1848, 2004.

**Saa, P., Castilla, J., and Soto, C., Cyclic amplification of protein misfolding and aggregation, in *Amyloid Proteins: Methods and Protocols*, Sigurdsson, E.M., Ed., Humana Press, Totowa, NJ, 2004, pp. 53–65.

*Saborio, G.P., Permanne, B., and Soto, C., Sensitive detection of pathological prion protein by cyclic amplification of protein misfolding, *Nature*, 411, 810–813, 2001. (The first report of the PMCA technology.)

Saborio, G.P. et al., Cell-lysate conversion of prion protein into its protease-resistant isoform suggests the participation of a cellular chaperone, *Biochem. Biophys. Res. Commun.*, 258, 470–475, 1999.

Schiermeier, Q., Testing times for BSE, *Nature*, 409, 658–659, 2001.

Soto, C., Altering prion replication for therapy and diagnosis of transmissible spongiform encephalopathies, *Biochem. Soc. Trans.*, 30, 569–574, 2002.

Soto, C., Diagnosing prion diseases: needs, challenges and hopes, *Nat. Rev. Microbiol.*, 2, 809–819, 2004.

*Soto, C. et al., Pre-symptomatic detection of prions by cyclic amplification of protein misfolding, *FEBS Lett.*, 579, 638–642, 2005. (Reports the application of PMCA for presymptomatic detection of PrP^{Sc} in the brain.)

Soto, C. and Castilla, J., The controversial protein-only hypothesis of prion propagation, *Nat. Med.*, 10, S63–S67, 2004

Soto, C. and Saborio, G.P., Prions: disease propagation and disease therapy by conformational transmission, *Trends Mol. Med.*, 7, 109–114, 2001.

**Soto, C., Saborio, G.P., and Anderes, L., Cyclic amplification of protein misfolding: application to prion-related disorders and beyond, *Trends Neurosci.*, 25, 390–394, 2002.

Telling, G.C. et al., Evidence for the conformation of the pathologic isoform of the prion protein enciphering and propagating prion diversity, *Science*, 274, 2079–2082, 1996.

Telling, G.C. et al., Prion propagation in mice expressing human and chimeric PrP transgenes implicates the interaction of cellular PrP with another protein, *Cell*, 83, 79–90, 1995.

Vey, M. et al., Subcellular colocalization of the cellular and scrapie prion proteins in caveolae-like membranous domains, *Proc. Natl. Acad. Sci. USA*, 93, 14945–14949, 1996.

Wadsworth, J.D. et al., Tissue distribution of protease resistant prion protein in variant Creutzfeldt-Jakob disease using a highly sensitive immunoblotting assay, *Lancet*, 358, 171–180, 2001.

chapter eleven

Other diseases of protein misfolding

The hallmark event in transmissible spongiform encephalopathies (TSEs) is the misfolding and aggregation of the prion protein (PrP) into a β-sheet-rich oligomeric structure that, by a not-yet-completely understood mechanism, produces brain degeneration. Misfolding and aggregation of PrP is also the basis by which the prion infectious agent (PrP^{Sc}) replicates in the body. Strikingly, a very similar molecular event is the hallmark process in a variety of human disorders, including the most prevalent neurodegenerative diseases and several systemic amyloidoses [Kelly, 1996; Carrell and Gooptu, 1998; Soto, 2001; Dobson, 2001]. These diseases, grouped under the name of protein misfolding disorders (PMD), are thought to arise from the misfolding and aggregation of an underlying protein. Recent findings strongly support this hypothesis and have increased our understanding of the molecular mechanism of PMD [Soto, 2001; Soto, 2003]. This chapter describes the clinical, etiological, and pathological features of these diseases, overviews the evidence in favor of protein misfolding as the key event in the disease pathogenesis, and discusses the molecular and structural basis underlying this process.

11.1 Protein misfolding and disease

Life depends on the correct function of a network of thousands of proteins that are essential for the proper activity of cells and organisms. The biological function of a protein depends on its three-dimensional structure, which is determined by its amino-acid sequence during the protein-folding process. This process is carefully supervised by chaperone proteins to avoid mistakes and to remove misfolded proteins [Fink, 1999]. However, under certain conditions protein misfolding and aggregation occur, leading to diverse diseases. In addition to TSEs, this group includes Alzheimer's disease (AD), serpin-deficiency disorders, hemolytic anemia, Huntington's disease (HD), cystic fibrosis, diabetes type II, amyotrophic lateral sclerosis (ALS),

Table 11.1 List of Some Diseases Classified in the Group of Protein
Conformational Disorders

Protein	Disease
Amyloid-β	Alzheimer's disease, Dutch hereditary amyloidosis
Tau	Alzheimer's disease, frontotemporal dementia
Amylin (islet amyloid polypeptide)	Diabetes type II
Prion protein	Prion diseases (spongiform encephalopathies)
α-Synuclein	Parkinson's disease, diffuse Lewy body disease
Huntingtin	Huntington's disease
Superoxide dismutase	Amyotrophic lateral sclerosis
Serum amyloid protein A	Secondary systemic amyloidosis
Transthyretin	Familial amyloid polyneuropathy type I
Apolipoprotein A-I	Familial amyloid polyneuropathy type II
Apolipoprotein A-II	Familial amyloid polyneuropathy type III
Apolipoprotein A-IV	Sporadic amyloid polyneuropathy
Lactoferrin	Corneal amyloidosis
Fibrinogen α-chain	Fibrinogen amyloidosis
Lysozyme	Lysozyme amyloidosis
Keratin	Cutaneous amyloidosis
Calcitonin	Medullary carcinoma of the thyroid
Prolactin	Aging pituitary prolactinomas
Insulin	Insulin-related amyloid
Atrial natriuretic factor	Atrial amyloidosis
Gelsolin	Finnish hereditary amyloidosis
Cystatin C	Icelandic hereditary amyloidosis
Serpins	Serpin deficiency, emphysema, cirrhosis
Immunoglobulin heavy or light chains	Primary systemic amyloidosis
β$_2$-Microglobulin	Hemodialysis-related amyloidosis
Kerato-epithelin	Corneal dystrophy
Cystic fibrosis transmembrane regulator	Cystic fibrosis
Androgen receptor	Spinal and bulbar muscular atrophy
Amyloid British	British familial dementia
Amyloid Danish	Danish familial dementia
Lactadherin (medin)	Aortic medial amyloidosis
Ataxins	Spinocerebellar ataxias
Atrophin I	Dentatorubropallidoluysian atrophy

secondary amyloidosis, Parkinson's disease (PD), dialysis-related amyloido-sis, and more than 15 other less-well-known diseases (Table 11.1).

The hallmark event in PMD is a change in the secondary and tertiary structure of a normal protein without alteration of the primary structure. The conformational change may promote the disease by either the gain of a toxic activity or by the loss of the biological function of the natively folded protein [Carrell and Gooptu, 1998; Soto, 2003]. The proteins implicated in each of the PMDs are different (Table 11.1), and there is no evident sequence or structural homology among the various proteins. However, the striking feature of these proteins is their inherent ability to adopt at least two different stable conformations [Carrell and Gooptu, 1998; Soto, 2001]. In most of the PMDs, the misfolded protein is rich in β-sheet conformation. β-Sheets are formed of alternating peptide-pleated strands linked by hydrogen bonding between the NH and CO groups of the peptide bond. Because the hydrogen bonding in the β-sheet can be either intramolecular or intermolecular, β-sheets are the preferred structures of protein aggregates.

Neuropathological, biochemical, and genetic studies as well as the development of transgenic animal models have provided strong evidence for the involvement of protein misfolding in disease. The first clue came several decades ago from postmortem histopathological studies. With the exception of cystic fibrosis, serpin deficiency disorders, and some forms of TSE, the end point of protein misfolding in PMD is aberrant protein aggregation and accumulation as amyloid-like deposits in diverse organs (Figure 11.1) [Glenner, 1980; Sipe, 1992; Thomas et al., 1995; Johnson, 2000]. Amyloid is the name originally given to extracellular protein deposits found in AD and systemic amyloid disorders, but it has recently been used to refer also to some of the intracellular aggregates. In this chapter, I use the term "amyloid-like" deposits to refer to these aggregates without necessarily meaning that they are absolutely equivalent. "Amyloid" is a generic term that refers to aggregates organized in a cross-β structure that contains specific morphological, tinctorial, and structural characteristics, such as binding to Congo red and thioflavin S, higher resistance to proteolytic degradation, and a fibrillar appearance under electron microscopy (straight, unbranched, 10-nm-wide fibrils) (Figure 11.2) [Glenner, 1990; Sipe and Cohen, 2000].

The correlation and colocalization of protein aggregates with degenerating tissue and disease symptoms is a strong indication of the involvement of amyloid deposition in the pathogenesis of PMD [Sipe, 1992; Ghiso et al., 1994; Soto, 2001]. Moreover, protein deposits have become a typical signature of PMD, and their presence is used for definitive diagnosis [Westermark, 1995; Gillmore et al., 1997]. However, it is still a matter of controversy whether the deposits of aggregated protein are the culprit of the disease or an inseparable epiphenomenon [Carrell and Gooptu, 1998; Tran and Miller, 1999; Goldberg and Lansbury, Jr., 2000; Soto et al., 2000; Caughey and Lansbury, Jr., 2003]. Postmortem studies showing, in some diseases, a poor correlation between the load of amyloid-like deposits and the severity of clinical symptoms [Terry et al., 1991] are often used to argue against a primary role of protein aggregates in PMD. Moreover, a mixture of different types of aggregates in some diseases and the appearance of protein deposits in clinically normal people [Terry, 1986; Hardy, 1997; Hamilton, 2000; Lee et al., 2004] further complicate a simple link between protein aggregation and disease.

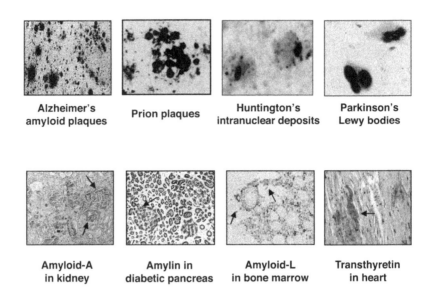

| Alzheimer's amyloid plaques | Prion plaques | Huntington's intranuclear deposits | Parkinson's Lewy bodies |

| Amyloid-A in kidney | Amylin in diabetic pancreas | Amyloid-L in bone marrow | Transthyretin in heart |

Figure 11.1 (See color insert after p. 114.) Histological staining of protein deposits in different tissues. Several distinct proteins have the ability to misfold and accumulate as amyloid-like plaques in different organs, triggering tissue damage and organ dysfunction.

Perhaps the most compelling evidence for the role of protein misfolding in disease comes from genetic studies [Buxbaum, 1996; Price et al., 1998; Hardy, 2001; Selkoe and Podlisny, 2002]. Most PMDs have both an inherited and sporadic origin. Interestingly, mutations in the genes encoding the protein component of fibrillar aggregates are genetically associated with inherited forms of the disease. The familial forms usually have an earlier onset and greater severity than sporadic cases. In the familial cases, a mutation seems to destabilize the normal protein folding, favoring the misfolding and aggregation of the protein. Mutations in the respective fibrillar proteins have been associated with familial forms of many diseases, including AD, TSE, HD and related polyglutamine disorders, PD, amyloid polyneuropathy, diabetes type II, cardiac amyloidosis, visceral amyloidosis, cerebral hemorrhage with amyloidosis of the Dutch and Icelandic type, cerebral amyloidosis of the British and Danish type, thromboembolic disease, angioedema, emphysema, sickle cell anemia, and ALS [Jacobson and Buxbaum, 1991; Kelly, 1996; Selkoe, 1996; Prusiner and Scott, 1997; Buxbaum and Tagoe, 2000]. Moreover, other genes genetically associated with the disease have also been shown to increase aggregation of the protein or decrease clearance of misfolded aggregates such as, for example, the presenilin genes in Alzheimer's disease that, when mutated, increase the production of the amyloid-β protein [Selkoe and Podlisny, 2002].

Another piece of evidence comes from the generation of transgenic animal models by the introduction of mutant forms of the human genes encoding the fibrillar protein. Several pathological and clinical features of diverse

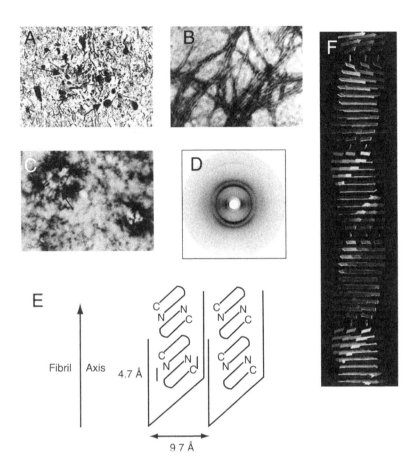

Figure 11.2 (See color insert after p. 114.) Morphological, tinctorial, and structural characteristics of amyloid aggregates. (A) A tissue staining with antibodies specific for the protein forming the aggregates reveals the accumulation of many punctuated aggregates. The localization of this protein deposit differs depending on the protein component and the particular disease. (B) Electron microscopy shows typical unbranched fibrils, which are 5 to 10 nm in diameter and 100 nm in length. (C) Cong red staining visualized under polarized light shows green/yellow birefringency, which is considered to be a typical signature for amyloid aggregates. (D) X-ray fiber diffraction studies show a characteristic pattern known as cross-β, represented by signals at 4.7 and 9.7 Å. (E) A diagrammatic picture for the cross-β structure showing the periodicity observed in X-ray diffraction studies. (F) A computer model of the putative tridimensional structure of a prototype amyloid fibril.

PMD have been observed in transgenic models in which protein aggregates were successfully produced [Araki et al., 1994; Janson et al., 1996; Price et al., 1998; Weissmann et al., 1998; Price et al., 2000; Gurney, 2000; Emilien et al., 2000].

Transgenic mice that overexpress high levels of human amyloid-precursor protein (APP) containing diverse mutations progressively develop many of the pathological hallmarks of AD, including cerebral amyloid deposits, neuritic dystrophy, astrogliosis, and cognitive and behavioral alterations [Price et al., 1998; Duff, 1998; Van Leuven, 2000; Emilien et al., 2000]. Transgenic mice expressing the wild-type human α-synuclein gene develop several of the clinicopathological features of PD, including accumulation of Lewy bodies in neurons of the neocortex, hippocampus, and substantia nigra; loss of dopaminergic terminals in the basal ganglia; and associated motor impairments [Masliah et al., 2000]. ALS pathology has been produced in mice by overexpressing the human mutated superoxide dismutase (SOD) gene [Price et al., 1997; Price et al., 1998]. Some of these mice develop motor neuron dysfunction similar to ALS patients as well as typical pathological alterations, including the presence of hyaline-inclusion bodies in degenerating axons, muscle atrophy and wasting, astrocytic damage, and extensive loss of large myelinated axons of motor neuronal cells. Transgenic mice containing the exon 1 of the human huntingtin and carrying 115 to 156 CAG repeat expansions develop pronounced neuronal intranuclear inclusions, containing the proteins huntingtin and ubiquitin, prior to developing a neurological phenotype [Scherzinger et al., 1997]. The cerebral abnormalities and clinical signs were very similar to those observed in HD patients. One of the first transgenic models showing a neurodegenerative process similar to a human disease was made by overexpression of the human mutated prion protein (PrP) gene [Hsiao et al., 1990; Weissmann et al., 1998]. Spontaneous neurological disease with spongiform degeneration was observed. Finally, a transgenic mouse model with high rates of expression of human islet amyloid polypeptide (IAPP) spontaneously developed diabetes mellitus by 8 weeks of age, which was associated with selective β-cell death and impaired insulin secretion [Janson et al., 1996]. Small intra- and extracellular IAPP aggregates were present in islets of transgenic mice during the development of diabetes mellitus.

Although these animal models supports a critical role for protein misfolding and polymerization in the disease, temporal studies of the appearance of diseaselike features in some of the models have shown that significant tissue damage and clinical symptoms appear before detection of protein aggregates [Janson et al., 1996; Moechars et al., 1999; Van Leuven, 2000]. These findings are usually interpreted as evidence for a misfolded soluble intermediate, not yet deposited in the tissue, as the real culprit of PMD [Carrell and Gooptu, 1998; Tran and Miller, 1999; Goldberg and Lansbury, Jr., 2000; Soto et al., 2000; Caughey and Lansbury, Jr., 2003].

11.2   Structural determinants of misfolding and aggregation

For most of the proteins implicated in PMD, many characteristics of the structural rearrangements featuring the pathogenic process have been modeled *in vitro*. Low-resolution structural studies have shown, in most of the cases (except for transthyretin, SOD1, immunoglobulin light chain, and serpins), a large secondary structural difference between the monomeric native protein and the aggregated material [Barrow et al., 1992; Pan et al., 1993; Conway et al., 2000; Chen et al., 2002]. In general, the misfolded protein is rich in β-sheet conformation, whereas the soluble normal protein exhibits diverse conformation, depending on the particular protein. The insolubility and noncrystalline nature of aggregated proteins has impeded high-resolution studies using conventional methods. However, studies using X-ray fiber diffraction and solid-state NMR have confirmed the β-sheet-rich structure of protein aggregates implicated in PMD [Serpell et al., 2000a; Serpell et al., 2000b]. An exception appears to be the structure of tau aggregates, which has recently been shown to be composed mainly of α-helices by studies using circular dichroism and Fourier-transformed infrared spectroscopy [Sadqi et al., 2002].

These studies have resulted in a molecular model of amyloid-like fibrils composed of several protofilaments consisting of hydrogen-bonding β-sheet structures, with the β-strands running perpendicular to the long-fiber axis, a structure known as a cross-β conformation (Figure 11.2). It is clear from these structural studies that a large conformational rearrangement of the polypeptide chain occurs during misfolding and aggregation. However, it is unknown whether the misfolding triggers protein aggregation or whether protein oligomerization induces the conformational changes [Soto, 2001]. On the basis of the available evidence, it is likely that slight conformational changes result in the formation of a misfolded intermediate, which is unstable in an aqueous environment and becomes stabilized by intermolecular interactions with other molecules, forming small β-sheet oligomers that, by further growth, produce amyloid-like fibrils (Figure 11.2). In this model, the conversion of the folded protein into the pathological form is triggered by structural changes, but complete misfolding depends on oligomerization.

11.3 Mechanism and driving forces in protein misfolding and aggregation

It is clear that all proteins in PMD undergo structural rearrangements during misfolding and aggregation. However, depending on the extent of the conformational changes and the thermodynamic forces driving the process, the diverse proteins can be classified in three different groups:

1. *Proteins in which misfolding and aggregation result in a profound structural rearrangement of the polypeptide chain driven by hydrophobic interactions.* The prototype proteins in this group are the Alzheimer's amyloid-beta (Aβ) peptide, Parkinson's α-synuclein, and the prion protein implicated in TSE. In these cases, the native protein adopts an α-helical or unstructured conformation, and aggregation is the

result of exposing the hydrophobic fragments to the aqueous solvent. Among this group of proteins, the Aβ peptide is undoubtedly the most extensively studied [Soto et al., 1994; Teplow, 1998]. Peptides containing the 40- or 42-residue forms of Aβ and shorter derivatives form amyloid-like fibrils *in vitro*, which are morphologically, tinctorially, immunologically, spectroscopically, and ultrastructurally similar to fibrillar aggregates extracted from AD amyloid plaques [Castano et al., 1986; Hilbich et al., 1992; Jarrett et al., 1993; Soto et al., 1995]. Studies with shorter Aβ fragments or with mutated peptides have shown that the internal hydrophobic region between amino acids 17 and 21 is the most important for the early steps of Aβ misfolding and aggregation, providing evidence for the idea that Aβ assembly is driven by hydrophobic interactions [Hilbich et al., 1992; Wood et al., 1995; Soto et al., 1995; Tjernberg et al., 1996]. This idea is consistent with the higher ability of Aβ peptides to aggregate with two or three additional hydrophobic amino acids at the C-terminal end [Jarrett et al., 1993]. Although less is known about the fibrilogenesis process of α-synuclein, the evidence suggests that the N-terminal fragment, 1 through 87, might be crucial for the misfolding and aggregation of the protein [Volles and Lansbury, Jr., 2003]. This fragment contains the hydrophobic NAC (nonamyloid component) peptide that was previously identified as a component of AD amyloid-like plaques by Saitoh and coworkers and was shown to be amyloidogenic [Iwai et al., 1995].

2. *Proteins in which misfolding and aggregation result in small changes in the secondary structure in a process driven by β-sheet instability.* The prototype members of this group are transthyretin involved in systemic amyloidosis, SOD1 implicated in ALS, and immunoglobulin light chain associated with primary amyloidosis (Table 11.1). The common characteristic among these proteins is that the native folding is rich in β-sheet, and during aggregation the intramolecular β-sheet interactions are replaced by intermolecular β-sheets [Kelly et al., 1997; Wall et al., 2004]. Aggregation is dependent upon destabilization of the native β-sheets, usually by changes in the quaternary structure or by mutations [Kelly et al., 1997; DiDonato et al., 2003]. Both in transthyretin and SOD1, protein polymerization is triggered by the conversion of the native tetramer or dimer into the monomer, in which the β-sheet packing becomes unstable and is prone to aggregate [Quintas et al., 2001; Elam et al., 2003; Hurshman et al., 2004]. One characteristic of this group is that the spreading of many different mutations throughout the protein sequence can lead to aggregation, most likely by destabilization of the native oligomers and conversion into the amyloidogenic monomers [Damas and Saraiva, 2000; Saraiva, 2001].

3. *Proteins in which misfolding and aggregation depend on glutamine/aspar-
    agine-rich domains and where protein aggregation is driven by hydrogen
    bonding involving both the peptide bonds and the side chains.* The proteins
    in this group are implicated in several human diseases in which
    mutations result in an abnormal and inherited expansion of CAG
    (codon codifying for glutamine) repeats [Zoghbi and Orr, 2000]. Some
    of the proteins in this group include huntingtin implicated in HD,
    ataxins associated with SCA, and the yeast prions. Aggregation of
    huntingtin and ataxins *in vitro* depends on the length of the poly-
    glutamine repeat [Scherzinger et al., 1997]. Glutamine and aspar-
    agine, while uncharged, have an amide group that provides a polar
    side chain and the possibility for hydrogen bonding. Perutz proposed
    a model named "polar zipper" [Perutz et al., 1994] to explain the
    cooperative interactions leading to β-sheet formation stabilized by
    double hydrogen bonding between the carbonyl and amide groups
    of the peptide bonds and the side chains of glutamine and asparagine.

11.4 Kinetics and intermediates of misfolding and aggregation

Kinetic studies have shown that aggregation of Aβ, huntingtin, α-synuclein,
amylin, and several other proteins follows a seeding/nucleation mechanism
[Jarrett et al., 1993; Wood et al., 1999; Scherzinger et al., 1999] that resembles
a crystallization process. The critical event is the formation of protein oligomers that act as a nucleus to direct further growth of aggregates. Nucleation-dependent polymerization is characterized by a slow lag phase in
which a series of unfavorable interactions result in the formation of an
oligomeric nucleus, which then rapidly grows to form larger polymers. The
lag phase can be minimized or removed by addition of preformed nuclei or
seeds [Harper and Lansbury, Jr., 1997].

At least two intermediates have been identified in the pathway from the
native monomeric protein to the fully aggregated fibrillar structure *in vitro*
(Figure 11.3). The first intermediate is the soluble, low-molecular-weight
oligomers (dimers to decamers) that have been identified in test-tube experiments, in the conditioned medium of cells that constitutively secrete the
protein, in human cerebrospinal fluid, and in human brain homogenate
[Levine, III, 1995; Kuo et al., 1996; Lambert et al., 1998; Walsh et al., 2002].
Structural and biochemical characterization of these intermediates has been
challenging because they are transient and unstable. However, it has been
proposed that some of these small oligomers may form pores in membranes,
destabilizing the homeostasis of essential metabolites and thus leading to
cell damage [Caughey and Lansbury, Jr., 2003].

The second intermediate corresponds to short, flexible, rodlike structures
called protofibrils (Figure 11.3), which have been studied by electron microscopy, photon correlation spectroscopy, and atomic force microscopy [Walsh

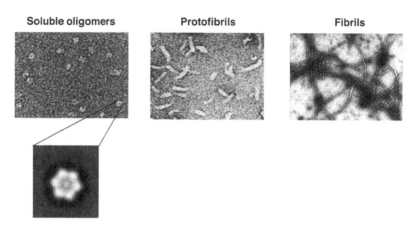

Figure 11.3 Intermediates in the process of fibril formation. At least two different structural intermediates in the progression from monomeric protein to the fibrillar state have been identified: soluble oligomers and protofibrils. The figure shows pictures of these structures as obtained by negative-stained electron microscopy.

et al., 1997]. Protofibrils are unbranched polymers 3- to 6-nm wide and up to 100-nm long. Kinetics studies have shown that they are metastable intermediates that elongate by coalescence of smaller protofibrils with a rate dependent on protein concentration, temperature, ionic strength, and pH of the medium [Harper et al., 1999]. Protofibrils are in a dynamic equilibrium with oligomeric protein and are the direct precursor of amyloid-like fibrils [Walsh et al., 1999]. Secondary-structure studies show that protofibrils have a high β-sheet content, like fibrils, and can bind the amyloid-specific dyes Congo red and thioflavin T [Walsh et al., 1999]. Evidence indicates that these intermediates, as well as the monomeric protein and the fibrillar aggregates, are all present simultaneously and are in dynamic equilibrium with each other [Levine, III, 1995; Teplow, 1998]. In addition, several lines of evidence suggest that low-molecular-weight oligomers and protofibrils can be even more toxic than mature amyloid-like fibrils, perhaps by making pores in the cellular membranes [Caughey and Lansbury, Jr., 2003; Klein et al., 2004; Walsh and Selkoe, 2004].

11.5   Interactions between misfolded proteins

Interestingly, in some PMDs, different types of protein aggregates are present simultaneously, and recent data suggest that the misfolded proteins may interact and even influence each other's aggregation processes [Hardy, 2003]. AD is the prototype disease in which two different forms of protein aggregates are present simultaneously in the brain: extracellular deposits of Aβ in senile amyloid plaques, and intracellular accumulations of neurofibrillary tangles composed of the hyperphosphorilated protein tau [Terry, 1994]. It has long been debated which of these two protein aggregates — amyloid plaques or neurofibrillary tangles — is mostly responsible for brain damage, and there is no consensus regarding the issue of whether the two aggregates interact with each other or act independently [Trojanowski, 2002]. A partial answer for both questions emerged from recent studies using transgenic mice bearing the human mutant APP and tau protein [Oddo et al., 2003]. These animals progressively developed both types of protein accumulations, and detailed longitudinal studies suggest that amyloid plaque deposition triggers tau aggregation.

Many AD patients develop signs of PD, and some PD patients become demented [Galasko et al., 1994]. Moreover, approximately 25% of patients with AD develop frank parkinsonism [Galasko et al., 1994], and Lewy-body-like inclusions develop in most cases of sporadic AD and familial AD [Lippa et al., 1999; Hamilton, 2000]. Moreover, Lewy bodies contain human Aβ, and Alzheimer's amyloid plaques contain fragments of α-synuclein [Van Gool et al., 1995; Iwai et al., 1995; Halliday et al., 1997]. Double transgenic mice bearing the mutant forms of the human APP and α-synuclein genes showed cognitive and motor alterations, loss of cholinergic neurons, synaptic alterations, extensive amyloid plaques, and Lewy-bodies-like intra-neuronal fibrillar inclusions [Masliah et al., 2001]. All of these features are also found in the Lewy-body variant of AD. These studies demonstrated that misfolding and aggregation of α-synuclein and Aβ have distinct, as well as convergent, pathogenic effects on the integrity and function of the brain. Expression of human α-synuclein did not affect the Aβ-dependent develop-ment of neuritic plaques or the overall Aβ content in the brain, but it wors-ened Aβ-dependent cognitive deficits and neurodegeneration in specific brain regions [Masliah et al., 2001]. These findings indicate that human α-synuclein may enhance the plaque-independent neurotoxicity of Aβ, which may help to explain the clinical observation that the Lewy-body variant of AD causes a more rapid cognitive decline than pure AD [Langlais et al., 1993]. On the other hand, overexpression of human APP promoted the intraneuronal accumulation of human α-synuclein and accelerated the devel-opment of motor deficits in transgenic mice.

A mechanistic explanation for the synergistic effect of diverse mis-folded protein aggregates comes from *in vitro* experiments in which aggre-gation of one protein was nucleated by oligomeric seeds prepared from another protein [Han et al., 1995]. This phenomenon is known as heter-ologous seeding, and its *in vivo* frequency and biological relevance remain to be studied.

11.6   Concluding remarks

Diverse human disorders, including several neurodegenerative diseases and systemic amyloidoses, are thought to arise from the misfolding and aggregation of an underlying protein. Several studies coming from differ-ent disciplines and distinct diseases strongly support this hypothesis and suggest that a common therapy for these devastating disorders might be possible. Many questions remain to be answered, but the implications of future research in this field are enormous. Despite the diversity of the proteins involved in PMD — proteins that do not share any sequential, structural, or functional similarities — the structure, morphology, and the mechanism of formation of misfolded protein aggregates are strikingly similar.

References * **

Araki, S. et al., Systemic amyloidosis in transgenic mice carrying the human mutant transthyretin (Met 30) gene: pathological and immunohistochemical similarity to human familial amyloidotic polyneuropathy, type I, *Mol. Neurobiol.*, 8, 15–23, 1994.

Barrow, C.J., Yasuda, A., Kenny, P.T., and Zagorski, M.G., Solution conformations and aggregational properties of synthetic amyloid beta-peptides of Alzheimer's disease: Analysis of circular dichroism spectra, *J. Mol. Biol.*, 225, 1075–1093, 1992.

Buxbaum, J., The amyloidoses, *Mt. Sinai J. Med.*, 63, 16–23, 1996.

Buxbaum, J.N. and Tagoe, C.E., The genetics of the amyloidoses, *Annu. Rev. Med.*, 51, 543–569, 2000.

**Carrell, R.W. and Gooptu, B., Conformational changes and disease — serpins, prions and Alzheimer's, *Curr. Opin. Struct. Biol.*, 8, 799–809, 1998.

*Castano, E.M. et al., *In vitro* formation of amyloid fibrils from two synthetic peptides of different lengths homologous to Alzheimer's disease beta-protein, *Biochem. Biophys. Res. Commun.*, 141, 782–789, 1986. (One of the first reports of amyloid fibril formation *in vitro* by a peptide associated with a PMD.)

**Caughey, B. and Lansbury, P.T., Jr., Protofibrils, pores, fibrils, and neurodegeneration: separating the responsible protein aggregates from the innocent bystanders, *Annu. Rev. Neurosci.*, 26, 267–298, 2003.

Chen, S. et al., Amyloid-like features of polyglutamine aggregates and their assembly kinetics, *Biochemistry*, 41, 7391–7399, 2002.

Conway, K.A., Harper, J.D., and Lansbury, P.T., Jr., Fibrils formed *in vitro* from alpha-synuclein and two mutant forms linked to Parkinson's disease are typical amyloid, *Biochemistry*, 39, 2552–2563, 2000.

Damas, A.M. and Saraiva, M.J., Review: TTR amyloidosis-structural features leading to protein aggregation and their implications on therapeutic strategies, *J. Struct. Biol.*, 130, 290–299, 2000.

DiDonato, M. et al., ALS mutants of human superoxide dismutase form fibrous aggregates via framework destabilization, *J. Mol. Biol.*, 332, 601–615, 2003.

**Dobson, C.M., Protein folding and its links with human disease, *Biochem. Soc. Symp.*, 1–26, 2001.

Duff, K., Transgenic models for Alzheimer's disease, *Neuropathol. Appl. Neurobiol.*, 24, 101–103, 1998.

Elam, J.S. et al., Amyloid-like filaments and water-filled nanotubes formed by SOD1 mutant proteins linked to familial ALS, *Nat. Struct. Biol.*, 10, 461–467, 2003.

---

* Highlights primary articles of outstanding importance and quality, including a short description of the findings.
** Highlights comprehensive review articles similar to the topic of this chapter.

Emilien, G., Maloteaux, J.M., Beyreuther, K., and Masters, C.L., Alzheimer disease: mouse models pave the way for therapeutic opportunities, *Arch. Neurol.*, 57, 176–181, 2000.

Fink, A.L., Chaperone-mediated protein folding, *Physiol. Rev.*, 79, 425–449, 1999.

Galasko, D. et al., Clinical-neuropathological correlations in Alzheimer's disease and related dementias, *Arch. Neurol.*, 51, 888–895, 1994.

"Ghiso, J., Wisniewski, T., and Frangione, B., Unifying features of systemic and cerebral amyloidosis, *Mol. Neurobiol.*, 8, 49–64, 1994.

Gillmore, J.D., Hawkins, P.N., and Pepys, M.B., Amyloidosis: a review of recent diagnostic and therapeutic developments, *Br. J. Haematol.*, 99, 245–256, 1997.

"Glenner, G.G., Amyloid deposits and amyloidosis: the beta-fibrilloses (first of two parts), *N. Engl. J. Med.*, 302, 1283–1292, 1980.

Glenner, G.G., The nomenclature of the cerebral amyloid fibril protein in Alzheimer's disease, *Neurobiol. Aging*, 11, 65, 1990.

Goldberg, M.S. and Lansbury, P.T., Jr., Is there a cause-and-effect relationship between alpha-synuclein fibrillization and Parkinson's disease? *Nat. Cell Biol.*, 2, E115–E119, 2000.

Gurney, M.E., What transgenic mice tell us about neurodegenerative disease, *Bioessays*, 22, 297–304, 2000.

Halliday, G. et al., Further evidence for an association between a mutation in the APP gene and Lewy body formation, *Neurosci. Lett.*, 227, 49–52, 1997.

Hamilton, R.L., Lewy bodies in Alzheimer's disease: a neuropathological review of 145 cases using alpha-synuclein immunohistochemistry, *Brain Pathol.*, 10, 378–384, 2000.

Han, H., Weinreb, P.H., and Lansbury, P.T., Jr., The core Alzheimer's peptide NAC forms amyloid fibrils which seed and are seeded by beta-amyloid: is NAC a common trigger or target in neurodegenerative disease? *Chem. Biol.*, 2, 163–169, 1995.

Hardy, J., The Alzheimer family of diseases: many etiologies, one pathogenesis? *Proc. Natl. Acad. Sci. USA*, 94, 2095–2097, 1997.

Hardy, J., Genetic dissection of primary neurodegenerative diseases, *Biochem. Soc. Symp.*, 51–57, 2001.

Hardy, J., The relationship between Lewy body disease, Parkinson's disease, and Alzheimer's disease, *Ann. N.Y. Acad. Sci.*, 991, 167–170, 2003.

Harper, J.D. and Lansbury, P.T., Jr., Models of amyloid seeding in Alzheimer's disease and scrapie: mechanistic truths and physiological consequences of the time-dependent solubility of amyloid proteins, *Annu. Rev. Biochem.*, 66, 385–407, 1997.

Harper, J.D., Wong, S.S., Lieber, C.M., and Lansbury, P.T., Jr., Assembly of A beta amyloid protofibrils: an *in vitro* model for a possible early event in Alzheimer's disease, *Biochemistry*, 38, 8972–8980, 1999.

Hilbich, C. et al., Substitutions of hydrophobic amino acids reduce the amyloidogenicity of Alzheimer's disease beta A4 peptides, *J. Mol. Biol.*, 228, 460–473, 1992.

Hsiao, K.K. et al., Spontaneous neurodegeneration in transgenic mice with mutant prion protein, *Science*, 250, 1587–1590, 1990.

Hurshman, A.R., White, J.T., Powers, E.T., and Kelly, J.W., Transthyretin aggregation under partially denaturing conditions is a downhill polymerization, *Biochemistry*, 43, 7365–7381, 2004.

Iwai, A. et al., The precursor protein of non-A beta component of Alzheimer's disease amyloid is a presynaptic protein of the central nervous system, *Neuron*, 14, 467–475, 1995.

Jacobson, D.R. and Buxbaum, J.N., Genetic aspects of amyloidosis, *Adv. Hum. Genet.*, 20, 69–11, 1991.

Janson, J. et al., Spontaneous diabetes mellitus in transgenic mice expressing human islet amyloid polypeptide, *Proc. Natl. Acad. Sci. USA*, 93, 7283–7288, 1996.

*Jarrett, J.T., Berger, E.P., and Lansbury, P.T., Jr., The carboxy terminus of the beta amyloid protein is critical for the seeding of amyloid formation: implications for the pathogenesis of Alzheimer's disease, *Biochemistry*, 32, 4693–4697, 1993. (One of the first reports describing the seeding nucleation model of amyloid formation.)

Johnson, W.G., Late-onset neurodegenerative diseases: the role of protein insolubility, *J. Anat.*, 196 (pt. 4), 609–616, 2000.

**Kelly, J.W., Alternative conformations of amyloidogenic proteins govern their behavior, *Curr. Opin. Struct. Biol.*, 6, 11–17, 1996.

Kelly, J.W. et al., Transthyretin quaternary and tertiary structural changes facilitate misassembly into amyloid, *Adv. Protein Chem.*, 50, 161–181, 1997.

Klein, W.L., Stine, W.B., Jr., and Teplow, D.B., Small assemblies of unmodified amyloid beta-protein are the proximate neurotoxin in Alzheimer's disease, *Neurobiol. Aging*, 25, 569–580, 2004.

Kuo, Y.M. et al., Water-soluble A beta (N-40, N-42) oligomers in normal and Alzheimer disease brains, *J. Biol. Chem.*, 271, 4077–4081, 1996.

Lambert, M.P. et al., Diffusible, nonfibrillar ligands derived from A beta 1-42 are potent central nervous system neurotoxins, *Proc. Natl. Acad. Sci. USA*, 95, 6448–6453, 1998.

Langlais, P.J. et al., Neurotransmitters in basal ganglia and cortex of Alzheimer's disease with and without Lewy bodies, *Neurology*, 43, 1927–1934, 1993.

Lee, V.M., Giasson, B.I., and Trojanowski, J.Q., More than just two peas in a pod: common amyloidogenic properties of tau and alpha-synuclein in neurodegenerative diseases, *Trends Neurosci.*, 27, 129–134, 2004.

Levine, H., III, Soluble multimeric Alzheimer beta(1-40) pre-amyloid complexes in dilute solution, *Neurobiol. Aging*, 16, 755–764, 1995.

Lippa, C.F., Schmidt, M.L., Lee, V.M., and Trojanowski, J.Q., Antibodies to alpha-synuclein detect Lewy bodies in many Down's syndrome brains with Alzheimer's disease, *Ann. Neurol.*, 45, 353–357, 1999.

Masliah, E. et al., Dopaminergic loss and inclusion body formation in alpha-synuclein mice: implications for neurodegenerative disorders, *Science*, 287, 1265–1269, 2000.

**Masliah, E. et al., Beta-amyloid peptides enhance alpha-synuclein accumulation and neuronal deficits in a transgenic mouse model linking Alzheimer's disease and Parkinson's disease, *Proc. Natl. Acad. Sci. USA*, 98, 12245–12250, 2001. (An interesting article showing *in vivo* evidence for interaction of different protein aggregates.)

Moechars, D. et al., Early phenotypic changes in transgenic mice that overexpress different mutants of amyloid precursor protein in brain, *J. Biol. Chem.*, 274, 6483–6492, 1999.

*Oddo, S. et al., Amyloid deposition precedes tangle formation in a triple transgenic model of Alzheimer's disease, *Neurobiol. Aging*, 24, 1063–1070, 2003. (This paper reports the relationship between amyloid plaque deposition and tangle formation in Alzheimer's disease using a triple transgenic mice model.)

Pan, K.M. et al., Conversion of alpha-helices into beta-sheets features in the formation of the scrapie prion proteins, *Proc. Natl. Acad. Sci. USA*, 90, 10962–10966, 1993.

Perutz, M.F., Johnson, T., Suzuki, M., and Finch, J.T., Glutamine repeats as polar zippers: their possible role in inherited neurodegenerative diseases, *Proc. Natl. Acad. Sci. USA*, 91, 5355–5358, 1994.

Price, D.L., Sisodia, S.S., and Borchelt, D.R., Genetic neurodegenerative diseases: the human illness and transgenic models, *Science*, 282, 1079–1083, 1998.

Price, D.L. et al., Amyotrophic lateral sclerosis and Alzheimer disease: lessons from model systems, *Rev. Neurol. (Paris)*, 153, 484–495, 1997.

Price, D.L. et al., The value of transgenic models for the study of neurodegenerative diseases, *Ann. N.Y. Acad. Sci.*, 920, 179–191, 2000.

Prusiner, S.B. and Scott, M.R., Genetics of prions, *Annu. Rev. Genet.*, 31, 139–175, 1997.

Quintas, A. et al., Tetramer dissociation and monomer partial unfolding precedes protofibril formation in amyloidogenic transthyretin variants, *J. Biol. Chem.*, 276, 27207–27213, 2001.

Sadqi, M. et al., Alpha-helix structure in Alzheimer's disease aggregates of tau-protein, *Biochemistry*, 41, 7150–7155, 2002.

Saraiva, M.J., Transthyretin amyloidosis: a tale of weak interactions, *FEBS Lett.*, 498, 201–203, 2001.

Scherzinger, E. et al., Huntington-encoded polyglutamine expansions form amyloid-like protein aggregates *in vitro* and *in vivo*, *Cell*, 90, 549–558, 1997.

Scherzinger, E. et al., Self-assembly of polyglutamine-containing Huntington fragments into amyloid-like fibrils: implications for Huntington's disease pathology, *Proc. Natl. Acad. Sci. USA*, 96, 4604–4609, 1999.

Selkoe, D.J., Cell biology of the beta-amyloid precursor protein and the genetics of Alzheimer's disease, *Cold Spring Harb. Symp. Quant. Biol.*, 61, 587–596, 1996.

Selkoe, D.J. and Podlisny, M.B., Deciphering the genetic basis of Alzheimer's disease, *Annu. Rev. Genomics Hum. Genet.*, 3, 67–99, 2002.

Serpell, L.C. et al., Fiber diffraction of synthetic alpha-synuclein filaments shows amyloid-like cross-beta conformation, *Proc. Natl. Acad. Sci. USA*, 97, 4897–4902, 2000a.

Serpell, L.C., Blake, C.C., and Fraser, P.E., Molecular structure of a fibrillar Alzheimer's A beta fragment, *Biochemistry*, 39, 13269–13275, 2000b.

*Sipe, J.D., Amyloidosis, *Annu. Rev. Biochem.*, 61, 947–975, 1992.

Sipe, J.D. and Cohen, A.S., Review: history of the amyloid fibril, *J. Struct. Biol.*, 130, 88–98, 2000.

*Soto, C., Protein misfolding and disease: protein refolding and therapy, *FEBS Lett.*, 498, 204–207, 2001.

*Soto, C., Unfolding the role of protein misfolding in neurodegenerative diseases, *Nat. Rev. Neurosci.*, 4, 49–60, 2003.

Soto, C., Branes, M.C., Alvarez, J., and Inestrosa, N.C., Structural determinants of the Alzheimer's amyloid beta-peptide, *J. Neurochem.*, 63, 1191–1198, 1994.

Soto, C., Castano, E.M., Frangione, B., and Inestrosa, N.C., The alpha-helical to beta-strand transition in the amino-terminal fragment of the amyloid beta-peptide modulates amyloid formation, *J. Biol. Chem.*, 270, 3063–3067, 1995.

Soto, C., Saborio, G.P., and Permanne, B., Inhibiting the conversion of soluble amyloid-beta peptide into abnormally folded amyloidogenic intermediates: relevance for Alzheimer's disease therapy, *Acta Neurol. Scand. Suppl.*, 176, 90–95, 2000.

Teplow, D.B., Structural and kinetic features of amyloid beta-protein fibrillogenesis, *Amyloid.*, 5, 121–142, 1998.

Terry, R.D., Interrelations among the lesions of normal and abnormal aging of the brain, *Prog. Brain Res.*, 70, 41–48, 1986.

Terry, R.D., Neuropathological changes in Alzheimer disease, *Prog. Brain Res.*, 101, 383–390, 1994.

Terry, R.D. et al., Physical basis of cognitive alterations in Alzheimer's disease: synapse loss is the major correlate of cognitive impairment, *Ann. Neurol.*, 30, 572–580, 1991.

**Thomas, P.J., Qu, B.-H., and Pedersen, P.L., Defective protein folding as a basis of human disease, *Trends Biochem. Sci.*, 20, 456–459, 1995.

Tjernberg, L.O. et al., Arrest of beta-amyloid fibril formation by a pentapeptide ligand, *J. Biol. Chem.*, 271, 8545–8548, 1996.

Tran, P.B. and Miller, R.J., Aggregates in neurodegenerative disease: crowds and power? *Trends Neurosci.*, 22, 194–197, 1999.

Trojanowski, J.Q., Taoists, Baptists, Syners, apostates, and new data, *Ann. Neurol.*, 52, 263–265, 2002.

Van Gool, D., De Strooper, B., Van Leuven, F., and Dom, R., Amyloid precursor protein accumulation in Lewy body dementia and Alzheimer's disease, *Dementia*, 6, 63–68, 1995.

Van Leuven, F., Single and multiple transgenic mice as models for Alzheimer's disease, *Prog. Neurobiol.*, 61, 305–312, 2000.

Volles, M.J. and Lansbury, P.T., Jr., Zeroing in on the pathogenic form of alpha-synuclein and its mechanism of neurotoxicity in Parkinson's disease, *Biochemistry*, 42, 7871–7878, 2003.

Wall, J.S. et al., Structural basis of light chain amyloidogenicity: comparison of the thermodynamic properties, fibrillogenic potential and tertiary structural features of four Vlambda6 proteins, *J. Mol. Recognit.*, 17, 323–331, 2004.

Walsh, D.M. et al., Amyloid beta-protein fibrillogenesis: structure and biological activity of protofibrillar intermediates, *J. Biol. Chem.*, 274, 25945–25952, 1999.

Walsh, D.M. et al., Naturally secreted oligomers of amyloid beta protein potently inhibit hippocampal long-term potentiation *in vivo*, *Nature*, 416, 535–539, 2002.

Walsh, D.M. et al., Amyloid beta-protein fibrillogenesis: detection of a protofibrillar intermediate, *J. Biol. Chem.*, 272, 22364–22372, 1997.

Walsh, D.M. and Selkoe, D.J., Oligomers on the brain: the emerging role of soluble protein aggregates in neurodegeneration, *Protein Pept. Lett.*, 11, 213–228, 2004.

Weissmann, C. et al., The use of transgenic mice in the investigation of transmissible spongiform encephalopathies, *Rev. Sci. Tech.*, 17, 278–290, 1998.

Westermark, P., Diagnosing amyloidosis, *Scand. J. Rheumatol.*, 24, 327–329, 1995.

Wood, S.J., Wetzel, R., Martin, J.D., and Hurle, M.R., Prolines and amyloidogenicity in fragments of the Alzheimer's peptide beta/A4, *Biochemistry*, 34, 724–730, 1995.

Wood, S.J. et al., Alpha-synuclein fibrillogenesis is nucleation-dependent: implications for the pathogenesis of Parkinson's disease, *J. Biol. Chem.*, 274, 19509–19512, 1999.

Zoghbi, H.Y. and Orr, H.T., Glutamine repeats and neurodegeneration, *Annu. Rev. Neurosci.*, 23, 217–247, 2000.

chapter twelve

Prions: a common phenomenon in biology?

Perhaps the most exciting research for the coming years will be focused on determining whether the prion phenomenon of propagation of biological information is the exclusive province of a limited group of proteins, like the prion protein (PrP), or rather is a more general process in biology. The findings of proteins with a prion-like behavior in yeast and other fungi have provided a step forward in this direction [Lindquist, 1997; Wickner et al., 1999]. Recent findings about the molecular mechanism by which prions replicate and propagate *in vivo* and *in vitro* suggest that other proteins forming β-sheet-rich amyloid-like aggregates have the possibility of behaving like prions. Exciting recent studies in other protein misfolding disorders (PMD) have provided good evidence for a potential infectious origin of other diseases of the group. This, coupled with the changing view that amyloids are a more general phenomenon and are not only associated with disease, raises the possibility that the prion phenomenon of propagating changes in protein function by transmission of alternative protein folding will be found to be much more common than we currently think.

12.1 The yeast prions

In 1994 Reed Wickner [Wickner, 1994] expanded the prion concept to explain the unusual non-Mendelian transmission of two yeast genetic elements termed [URE3] and [PSI$^+$], which were proposed to be the prion forms of the Ure2 and Sup35 proteins, respectively. Sup35p plays an essential role in translation termination, and Ure2p participates in the modulation of nitrogen metabolism (Figure 12.1) [Lindquist, 1997; Masison et al., 2000]. The prion phenotype in yeast is caused by protein aggregation and sequestration, leading to the loss of function of the native protein. For example, Sup35p, a component of the translation termination complex (Figure 12.1A), spontaneously aggregates at low frequency into a functionally impaired form, which then recruits all Sup35p molecules into the prion state [Glover et al.,

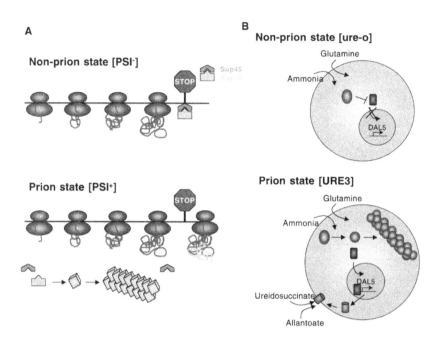

Figure 12.1 Schematic representation of the Sup35p and Ure2p yeast prions. (A) Sup35p plays an essential role in translation termination. In the nonprion state (termed [PSI⁻]), soluble Sup35 binds Sup45 to form a complex that binds to certain mRNA sequences that encode a translation stop site, resulting in termination of translation. In the prion state (termed [PSI⁺]), Sup35 is recruited in misfolded amyloid-like aggregates and thus is not available to stop translation, resulting in longer versions of certain proteins. (B) Ure2p participates in the modulation of nitrogen metabolism. In the nonprion state (termed [ure-o]), soluble Ure2p interacts with Gln3, preventing its translocation to the nucleus and its induction of the transcription of the DAL5 gene. In the prion state (termed [Ure3]), Ure2p is sequestered into amyloid fibrils and thus is not available to interact with Gln3, which translocates to the nucleus, triggering the transcription of the DAL5 gene. A DAL5 protein is inserted in the plasma membrane, thus enabling the internalization of ureidosuccinate and allantoate to the cell.

1997]. This state is passed on to the progeny of the [PSI⁺] cell. Over the course of the 10 years since the milestone paper from Wickner, extensive evidence has accumulated in favor of the prion hypothesis in yeast [Uptain and Lindquist, 2002; Wickner et al., 2004]. Moreover, several other proteins exhibiting the prion phenomenon were identified in yeast and other fungi [Uptain and Lindquist, 2002].

A yeast prion can be defined as an infectious protein that behaves as a non-Mendelian genetic element that transmits biological information in the absence of nucleic acid. In yeast, prions do not spread from cell to cell and do not kill the cells harboring them. They are inherited by daughter cells from their mothers and produce new metabolic phenotypes. However, like mammalian prions, they are based upon the ability of a protein that has acquired an abnormal conformation to influence other proteins of the same type to adopt the same conformation. Thus, in both cases, protein structures act in a manner previously thought to be unique to nucleic acids: in the mammalian case as transmissible agents of disease, and in yeasts as heritable determinants of phenotype.

Diverse genetic, biochemical, and structural evidence has been provided in support of the prion nature of the yeast determinants. (For references see [Uptain and Lindquist, 2002; Wickner et al., 2004].):

1. The non-Mendelian genetic element can be passed by cytoplasmic transference in the absence of nucleic acids.
2. The chromosomal gene encoding the normal form is required to propagate the prion form.
3. Overproduction of the normal protein increases spontaneous occurrence of the prion form.
4. The prion phenotype can be cured by protein denaturation, and it can arise again spontaneously at some low frequency.
5. Like PrP, the yeast prions can exist in two conformational states, a normal soluble and protease-sensitive state and an insoluble, protease-resistant β-sheet-rich aggregated form.
6. The conversion process has been reproduced *in vitro* using highly purified proteins that, when integrated in the cells, changed the yeast phenotype.
7. The replication process was repeated sequentially after serial dilutions, mimicking the continuous propagation of prions.
8. The prion-forming domain of Sup35p is modular and transferable, and indeed artificial prions have been generated by fusing a mammalian receptor to the yeast prion domain.

Although yeast prions were discovered much later than mammalian prions, the remarkably rapid progress in these studies has already provided some important implications for understanding the underlying biology of prions and the protein-only nature of the "infectious agent." The first breakthrough came from studies by Sparrer et al. [2000], showing that introduction of *in vitro*-converted purified Sup35 prion domain (N-terminal residues 1–254) via a liposome-based transformation procedure caused the appearance of [PSI+] prion in 1 to 2% of transformed cells. However, the low efficiency of this procedure did not allow ruling out the possibility that infectivity was produced *de novo* as a result of a high local concentration of Sup35p. A step forward in the direction of the prion hypothesis came from the studies by Maddelein et al. [2002], using the [Het-s] prions from *Podospora anserine*. These authors showed that insertion of fibrils made *in vitro* from renatured recombinant HET-s into the mycelia of *P. anserine* causes the efficient appearance of the [Het-s] prions.

A recent study [King and Diaz-Avalos, 2004] showed that bacterially produced N-terminal fragments of Sup35p labeled with green fluorescent protein, when transformed into amyloid fibrils by incubation with yeast-derived infectious aggregates, were able to propagate the prion phenotype to yeast cells [King and Diaz-Avalos, 2004]. *De novo* generation of infectivity was demonstrated in experiments where samples containing infectious particles were incubated with recombinant Sup35 fragments and subjected to sonication followed by serial dilutions into more recombinant protein. After three rounds of incubations, sonications, and dilutions, the sample contained a 160-fold dilution of the original infectious material. Interestingly, although the newly converted material retained infectivity, no activity was observed in the original seeding material at the same dilution [King and Diaz-Avalos, 2004]. Moreover, the *in vitro*-converted protein faithfully propagated the characteristics of the several strains used to begin the conversion reaction. These findings imply that the heritable information encoding distinct strains resides exclusively in the folding patterns of the protein.

The same conclusion was obtained independently by Weissman and coworkers [Tanaka et al., 2004]. Amyloid fibrils produced *de novo* from the recombinant Sup35 prion domain at different temperatures adopted distinct, stably propagating conformations, as characterized by thermal stability and electron paramagnetic resonance spectroscopy. Infection of yeast with these different amyloid conformations led to distinct [PSI⁺] strains *in vivo* [Tanaka et al., 2004]. In a follow-up study, the same group generated a special folding of Sup35p from *S. cerevisiae* that was capable of infecting this species as well as the highly divergent *C. albicans* both *in vivo* and *in vitro* [Tanaka et al., 2005]. Interestingly, the new strain generated in *C. albicans* conserved a memory to reinfect the *S. cerevisiae* species. These results support a model whereby strain conformation is the critical determinant of cross-species prion transmission, while primary structure affects transmission specificity by altering the spectrum of preferred amyloid conformations [Tanaka et al., 2005].

12.2 The inherent infectious nature of misfolded aggregates

As discussed in Chapter 11, protein misfolding and aggregation is a hallmark feature not only of prion diseases, but also of a group of more than 25 distinct human disorders. Normal cellular protein (PrPᶜ) converts into pathological prion protein (PrPˢᶜ) in transmissible spongiform encephalopathy (TSE). Similarly, the protein conformational changes associated with the pathogenesis of these diseases result in the formation of abnormal proteins rich in β-sheet structure, partially resistant to proteolysis, and with a high tendency to form larger-order aggregates [Carrell and Lomas, 1997; Dobson, 1999; Soto, 2001]. Indeed, a common feature of several protein misfolding diseases (PMD), including TSE, is the aggregation and deposition of the misfolded protein in different organs in the form of amyloid-like plaques. Interestingly, the available data indicate that amyloid formation in all cases follows a seeding-nucleation mechanism [Jarrett and Lansbury, Jr., 1993]. Analogous to a crystallization process, amyloid formation depends on the slow interaction between misfolded protein monomers to form oligomeric nuclei around which a faster phase of elongation takes place (Figure 12.2). Much like PrP^{Sc} in TSEs, the oligomeric nuclei act as a seed to induce and stabilize the conversion of the native monomeric protein (Figure 12.2). The limiting step in this process is the nuclei formation, and the extent of amyloidosis depends on the number of seeds produced [Jarrett and Lansbury, Jr., 1993; Lomakin et al., 1996]. These theoretical considerations have been extensively supported by experimental data, and it is widely accepted that protein misfolding and aggregation follow a seeding-nucleation mechanism.

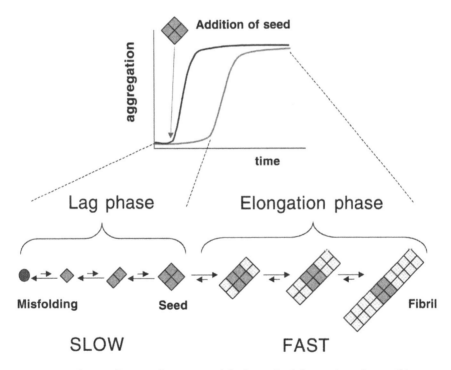

Figure 12.2 The seeding-nucleation model of amyloid formation. Compelling evidence suggests that amyloid fibril formation follows a crystallizationlike process consisting of lag and elongation phases. During the lag phase, oligomeric nuclei are formed in a slow process that involves misfolding of the protein and unfavorable intermolecular interactions. Once seeds are formed, a much more rapid phase of elongation results in fibril formation. The limiting step in the process is the formation of seeds to direct further aggregation. Amyloid formation can be substantially accelerated by addition of preformed seeds that represent the inherently infectious structure.

The seeding-nucleation model provides a rationale and plausible explanation for the infectious nature of prions (see Chapters 3 and 10) and suggests that protein misfolding processes, such as those associated with several human diseases, have the inherent ability to be transmissible. Infectivity lies in the capacity of preformed, stable, misfolded oligomeric proteins to act as seeds catalyzing the misfolding and aggregation process (Figure 12.2) [Gajdusek, 1994]. The acceleration of protein aggregation by addition of seeds has been convincingly reported *in vitro* for several unrelated proteins implicated in diverse PMDs. Extrapolating the *in vitro* results to the *in vivo* situation, the correct administration of a preaggregated stable misfolded structure should substantially enhance the misfolding, aggregation, and tissue accumulation of the protein. Provided that protein misfolding and aggregation are the cause of the disease, this should lead to the acceleration of a pathogenic process that, in the absence of the seed, was set to occur much later in life or not at all during the life span of the individual.

12.3   Why are protein misfolding disorders other than TSE not infectious?

Transmissibility of amyloidosis and other PMDs has not been thoroughly investigated [Sigurdsson et al., 2002], but it is generally assumed, based on epidemiological studies, that they do not have an infectious origin. For example, family members or medical professionals working with PMD patients do not have a higher propensity to develop the disease. However, the same is true for prion diseases. It must be emphasized that the principles that generally apply to conventional infectious diseases do not necessarily hold true for this protein-only agent, which follows complicated mechanism of transmission and requires special routes of infection. In addition, the long incubation times (up to several decades in humans) further complicate tracking a potentially infectious cause. The challenges to clearly establish an infectious origin will likely be greater in much more prevalent diseases, such as Alzheimer's or diabetes type II. How can we be sure that some of the cases of these common diseases do not have an infectious origin? We certainly cannot rule out this possibility! A possible infectious contribution in accelerating the disease onset might explain why, among people with similar genetic backgrounds (for example identical twins) and exposed to comparable environmental conditions, only some develop these devastating diseases while others stay free of the disease until death by other causes.

Perhaps the best way of investigating the infectious propagation of PMDs is to attempt to transmit the disease to experimental animals. Here lies a crucial difference between TSE and other PMDs: while TSEs naturally affect both humans and animals, most PMDs are restricted to humans. In addition, the experimental transgenic animal models for PMDs recapitulate only partially the characteristics of the human disease [Price et al., 1998], whereas the animal models of TSE perfectly reproduce all features of human prion disorders [Kimberlin, 1976]. Despite that, human prions cannot infect wild-type rodents because of the species barrier (see Chapter 6). Even using transgenic animals expressing the human gene, transmissibility from humans is not efficient unless the mouse gene is removed [Telling et al., 1994]. By extrapolating these findings to other PMDs, it is obvious that selection of the right experimental model is crucial to maximize the probability of successful transmission.

Several attempts have been made to transmit Alzheimer's disease (AD) to experimental animals, with intriguing but controversial results [Goudsmit et al., 1980; Brown et al., 1982; Wisniewski et al., 1984; Baker et al., 1993; Brown et al., 1994]. Marmosets injected with AD brain homogenates developed scattered deposits of the amyloid-β protein (Aβ) in the brain parenchyma and cerebral vasculature 6 to 7 years after inoculation [Baker et al., 1994]. Interestingly, the resultant amyloid lesions were not limited to the injection site. However, extensive studies by Gajdusek and colleagues failed to transmit AD and other dementias to primates [Brown et al., 1994]. It is important to note that, for most of these studies, a single primate was used for inoculation. Thus, it would be important to repeat some of these experiments using a larger cohort of animals for a definitive answer on transmissibility. More-recent studies have used transgenic mice expressing the human mutant amyloid precursor protein gene. Kane and coworkers infused diluted brain homogenates, derived from Alzheimer's patients after autopsy, unilaterally into the hippocampus and neocortex of 3-month-old transgenic mice [Kane et al., 2000; Walker et al., 2002]. Up to 4 weeks after infusion, there was no Aβ deposition in the brain; however, after 5 months, transgenic mice developed profuse Aβ-immunoreactive senile plaques and vascular deposits exclusively in the hemisphere injected [Kane et al., 2000]. After 12 months, abundant Aβ deposits were present bilaterally in the forebrain, but plaque load was still clearly greater in the injected hemisphere [Walker et al., 2002]. These findings clearly show that preformed Aβ aggregates can enhance *in vivo* plaque deposition. However, since these transgenic animals develop AD pathology "spontaneously" later on, it is not possible to determine whether inoculation with AD brain acted as an infectious agent or just as an accelerator of a process that was genetically programmed to occur.

Perhaps the best evidence for a prion-like phenomenon in other PMDs comes from the pioneering studies of the groups led by Westermark and Higuchi, who worked in systemic amyloidosis associated with deposition of amyloid-A (AA) and apolipoprotein AII amyloid (AApoAII), respectively.

AA is a fragment of serum amyloid-A (SAA) protein, which is an acute phase reactant apolipoprotein [Sipe et al., 1985; Rocken and Shakespeare, 2002]. The plasma concentration of SAA is normally low (around 20 mg/L), but can increase to >1000 mg/L as a result of an inflammatory stimulus [De Beer et al., 1982]. AA is a 76-residue cleavage product from the N-terminal region of SAA that is deposited systemically as amyloid in vital organs, including the liver, spleen, and kidneys [Sipe et al., 1985]. Clinically, AA amyloidosis occurs in patients with rheumatoid arthritis and other chronic inflammatory diseases [Rocken and Shakespeare, 2002]. The disease can be induced experimentally in mice by an inflammatory stimulus to dramatically increase SAA concentrations by injections of silver nitrate, casein, or lipopolysaccharide [Skinner et al., 1977]. Two to three weeks after the inflammatory stimulus, the animals develop systemic AA deposits like those found in patients with AA amyloidosis. This lag phase is dramatically shortened to a few days when mice are given, concomitantly, an IV injection of protein extracted from the spleen or liver containing amyloid plaques [Kisilevsky and Boudreau, 1983]. This tissue preparation is often referred to as amyloid enhancing factor (AEF) [Kisilevsky et al., 1999]. Although the nature of AEF has been extensively studied, recent experiments have demonstrated that the active principle in the AA amyloid extracts was the amyloid fibril itself [Ganowiak et al., 1994; Johan et al., 1998].

Interestingly, it was demonstrated that inoculation of animals with AEF led to the acceleration of the disease even at minuscule doses (<1 ng) and that AEF retained its biological activity over a considerable length of time [Lundmark et al., 2002]. Notably, the AEF was also effective when administered orally and could be serially transferred among animals. Furthermore, treatment of AEF with denaturing agents completely abrogated its activity [Lundmark et al., 2002]. These findings come tantalizingly close to the characteristics of infectious prion proteins. However, there are two important differences: (1) injection of AEF in the absence of an inflammatory stimulus does not lead to disease, and (2) amyloid deposition and disease appear after inflammation even in the absence of AEF inoculation. Therefore, analogous to the results in transgenic mice models of AD, the phenomenon cannot be classified as infectious but, rather, as just an acceleration of the disease process.

AApoAII accumulates in diverse organs during aging, leading to mouse senile amyloidosis [Higuchi et al., 1995]. Apolipoprotein AII (apoAII) is the second-most-abundant apolipoprotein in serum high-density lipoproteins (HDL) [Hatters and Howlett, 2002]. A single intravenous injection of a very small amount of the AApoAII fibrils induced severe systemic amyloid deposition in young mice [Higuchi et al., 1998]. After AApoAII injection, amyloid deposition occurred rapidly and advanced in an accelerated manner, as observed in spontaneous senile amyloidosis in mice. Strikingly, injection of denatured AApoAII fibrils, native nonfibrillar apoAII in HDL particles, or denatured apoAII monomer did not induce amyloidosis [Higuchi et al., 1998]. In a follow-up study, the same group showed that oral administration of AApoAII amyloid fibrils in mice for five consecutive days resulted in all animals developing amyloid deposits at 2 months of age. The plaques were located initially in the small intestine, but they extended to the tongue, stomach, heart, and liver at 3 to 4 months after feeding [Xing et al., 2001]. Interestingly, amyloid deposition was observed in young mice raised in the same cage for 3 months with old mice who had severe amyloidosis. Detection of AApoAII in foces of old mice and induction of amyloidosis by the injection of a fraction of feces suggested that propagation of amyloidosis between animals in the same cage was likely by eating feces [Xing et al., 2001].

Interestingly, transmission of AApoAII amyloidosis exhibits a "strain phenomenon" analogous to the prion strains. Three polymorphic variants of apoAII (types A, B, and C) with different amino acid substitutions at four positions are present in inbred strains of mice [Higuchi et al., 1991]. Senescence-accelerated mouse-resistant 1 (SAMR1) mice with wild-type apoAII of the type B develop few, if any, amyloid plaques spontaneously. Conversely, a congenic strain of mice termed R1.P1-*ApoAII* that expresses the allele C of apoAII gene spontaneously exhibited a high incidence of amyloid plaques with aging [Higuchi et al., 1995]. Injection of AApoAII fibrils from type C apolipoprotein AII (AApoAII-C) intravenously into 2-month-old SAMR1 mice (expressing type B ApoAII) induced extensive amyloid deposits in the tongue, stomach, intestine, lungs, heart, liver, and kidneys, as detected 10 months after injection [Xing et al., 2002]. The intensity of deposition increased thereafter, whereas no amyloid was detected in distilled-water-injected SAMR1 mice, even after 20 months. The deposited amyloid was composed of endogenous ApoAII type B (AApoAII-B) with a different amyloid fibril conformation than AApoAII-C [Xing et al., 2002]. Subsequent injection of these AApoAII-B amyloid fibrils induced earlier and more severe amyloidosis in SAMR1 mice than the injection of AApoAII-C fibrils [Xing et al., 2002]. These results suggest that inoculation of AApoAII-C into animals genetically immune to the disease can induce an infectious process, leading to the formation of a conformationally altered version of ApoAII-B to a different amyloid fibril structure, which could also induce amyloidosis in the less amyloidogenic strain. Putting all these findings together, there is a compelling case that some of the proteins associated with other amyloid-related disorders can have prion-like infectious properties. Whether or not some of these diseases have an infectious origin under nonexperimental conditions remains to be studied.

12.4   How common is the prion phenomenon in nature?

At present we do not know if the prion phenomenon operates in normal biological processes, but it seems reasonable that the efficient conversion of a protein function by transmission of an alternative folding provides an excellent way to modulate the activity of proteins without the need for genetic changes. As described previously, yeasts and other fungi discovered these advantages long ago during evolution. The crucial question is, how many more prions are constantly operating in living cells? Identification of new prions is a difficult task, especially when the putative result of the prion activity is not a dramatic disease as in TSE or a weird genetic property as in yeast prions, but just a slight change in the biological function of a particular protein. A very interesting recent study from Kandel and coworkers reported prion-like properties in a neuronal member of the CPEB family (cytoplasmic polyadenylation element binding protein), which regulates mRNA translation [Si et al., 2003]. The neuronal isoform of *Aplysia* CPEB is causally involved in maintaining long-term synaptic facilitation. They fused the putative prion domain (analogous to the yeast prion domain) to a reporter protein and found the epigenetic changes in state that characterize yeast prions [Si et al., 2003]. Full-length CPEB undergoes similar changes, but surprisingly it is the dominant, self-perpetuating prion-like form that has the greatest capacity to stimulate translation of CPEB-regulated mRNA. The authors hypothesized that conversion of CPEB to a prion-like state in stimulated synapses helps to maintain long-term synaptic changes associated with memory storage [Si et al., 2003]. The prion properties of CPEB were suspected because this protein has an asparagine/glutamine-rich domain, which in yeast has been demonstrated to constitute the responsible sequence for the prion activity. However, mammalian prions do not contain this type of sequence, and thus the search for asparagine/glutamine-rich domains is only one way to identify novel prions.

A common characteristic among mammalian, yeast, fungi, and aplysia prions is that the prion state is characterized by a particular physical state of the protein forming β-sheet-rich aggregates that resemble amyloid fibrils. Protein oligomerization in β-sheet structures is not only a typical characteristic of prions, but as discussed above, it is probably key for prions to propagate by seeding. Therefore, one way to identify novel "biological" prions is to search for proteins forming β-sheet amyloid-like aggregates as part of normal cellular functioning. The conception of amyloid as the product of an inherently pathological process associated with a small group of proteins has dramatically changed within the last few years. The pioneering work of Dobson and colleagues has demonstrated that many (if not all) proteins can make β-sheet intermolecular interactions to adopt an amyloid-like conformation under appropriate conditions [Stefani and Dobson, 2003]. The SH3 domain of PI3K, acylphosphatase, apomyoglobin, and fibronectin are among the growing list of proteins that have been made to form amyloid under experimental conditions [Stefani and Dobson, 2003]. Aggregation of these proteins to form amyloid fibrils was induced by changes on experimental solution conditions, such as low pH, high temperature, and moderate concentrations of salts or organic solvents such as trifluoroethanol. These results indicated that the ability to form amyloid fibrils is a generic property of peptides and proteins, and that protein aggregation was the result of the inherent physicochemical properties of the polypeptide chain rather than the specific interactions of side chains [Stefani and Dobson, 2003]. Importantly, the mechanism of aggregation in these "artificial" amyloids is the same as in the disease-associated amyloid structures, which are characterized by the seeding-nucleation model. Therefore, the finding that most (if not all) proteins can be folded into β-sheet-rich amyloid-like structures by a seeding mechanism suggests that every protein has, in principle, the possibility to act as a prion.

In addition, several proteins have recently been shown to aggregate naturally into amyloid-like structures, and the formation of these aggregates

Table 12.1 List of Proteins Forming Amyloid as Part of Their Normal Biological Function

Protein	Organism	Biological Function of Amyloid
Curli	Bacteria	Biofilm formation in *E. coli*
Microcin	Bacteria	Regulates the bacterial toxin activity of the protein
Hydrophobin	Fungi	Forms a water-resistant coat mesh
Fibroin and other silk proteins	Spiders	Networks of fibrils that constitute a spider web
Chorion	Insects and fish	Produces a dehydration-resistant shell
Pmel 17	Humans	Participates in melanin packaging by melanocytes

appears to be associated with a beneficial biological function [Kelly and Balch, 2003]. These proteins include (Table 12.1): curli, which is involved in biofilm formation in *E. coli* [Chapman et al., 2002]; hydrophobin, which acts as a water-resistant coat for fungi [Wosten and de Vocht, 2000]; chorion, which produces a dehydration-resistant shell on eggs of fish and insects [Iconomidou et al., 2000]; fibroin and other silk proteins that are used in the formation of spider webs [Kenney et al., 2002]; Pmel 17, which is involved in melanin packaging by human melanocytes [Berson et al., 2003]; and most recently, microcin, which is a bacterial toxin that, upon oligomerization, makes ionic channels in membranes of sensitive bacteria [Bieler et al., 2005]. In all these instances, protein aggregation in β-sheet amyloid structures is normally produced and is crucial for the biological activity of the protein. Therefore, it is clear from these examples that amyloids have been used and evolutionarily selected by diverse organisms to perform important biological functions. Again, in several of these proteins, it has been shown that amyloid formation follows a seeding-nucleation model [Li et al., 2001; Bieler et al., 2005]. Whether or not these "beneficial" amyloids can be propagated by a prion-like phenomenon has not been investigated, but it remains an intriguing possibility.

12.5   Concluding remarks

Future research should help us answer the question of whether the prion phenomenon of transmission of biological information by propagation of protein (mis)folding is a common biological process. The discovery of yeast prions, along with the fact that the molecular mechanism underlying prion propagation is strikingly similar to the mechanism of amyloid formation, suggests that prions may be much more common in nature than we currently think. In addition, the findings that amyloid is not restricted to diseases suggest that the prion phenomenon may play an important role in normal biological processes. It seems possible that transmission of protein conformation from one molecule to another may represent a natural process to modify, within one generation, the structure and activity of proteins to adapt to new conditions. The possibility that many proteins adopt multiple conformations to exert different functions, and that this biological information can be propagated between different individuals, might revolutionize our understanding of biology.

References* **

*Baker, H.F. et al., Experimental induction of beta-amyloid plaques and cerebral angiopathy in primates, *Ann. N.Y. Acad. Sci.*, 695, 228–231, 1993. (An intriguing study showing evidence for the transmission of Alzheimer's disease to primates.)

Baker, H.F. et al., Induction of beta (A4)-amyloid in primates by injection of Alzheimer's disease brain homogenate: comparison with transmission of spongiform encephalopathy, *Mol. Neurobiol.*, 8, 25–39, 1994.

Berson, J.F. et al., Proprotein convertase cleavage liberates a fibrillogenic fragment of a resident glycoprotein to initiate melanosome biogenesis, *J. Cell Biol.*, 161, 521–533, 2003.

Bieler, S., Amyloid formation modulates the biological activity of a bacterial protein, *J. Biol. Chem.*, 280, 26880–26885, 2005.

Brown, P. et al., Human spongiform encephalopathy: the National Institutes of Health series of 300 cases of experimentally transmitted disease, *Ann. Neurol.*, 35, 513–529, 1994.

Brown, P., Salazar, A.M., Gibbs, C.J., Jr., and Gajdusek, D.C., Alzheimer's disease and transmissible virus dementia (Creutzfeldt-Jakob disease), *Ann. N.Y. Acad. Sci.*, 396, 131–143 1982.

Carrell, R.W. and Lomas, D.A., Conformational disease, *Lancet*, 350, 134–138, 1997.

*Chapman, M.R. et al., Role of *Escherichia coli* curli operons in directing amyloid fiber formation, *Science*, 295, 851–855, 2002. (The first study reporting formation of amyloid in bacteria, which play a beneficial role in biofilm formation.)

De Beer, F.C., et al., Serum amyloid-A protein concentration in inflammatory diseases and its relationship to the incidence of reactive systemic amyloidosis, *Lancet*, 2, 231–234, 1982.

Dobson, C.M., Protein misfolding, evolution and disease, *Trends Biochem. Sci.*, 24, 329–332, 1999.

**Gajdusek, D.C., Nucleation of amyloidogenesis in infectious and noninfectious amyloidoses of brain, *Ann. N.Y. Acad. Sci.*, 724, 173–190, 1994.

*Ganowiak, K., et al., Fibrils from synthetic amyloid-related peptides enhance development of experimental AA-amyloidosis in mice, *Biochem. Biophys. Res. Commun.*, 199, 306–312, 1994. (One of the first reports showing acceleration of protein misfolding and aggregation in systemic amyloidosis by seeding with preformed fibrils *in vivo*.)

Glover, J.R. et al., Self-seeded fibers formed by Sup35, the protein determinant of [PSI+], a heritable prion-like factor of *S. cerevisiae*, *Cell*, 89, 811–819, 1997.

Goudsmit, J. et al., Evidence for and against the transmissibility of Alzheimer disease, *Neurology*, 30, 945–950, 1980.

---

* Highlights primary articles of outstanding importance and quality, including a short description of the findings.

** Highlights comprehensive review articles similar to the topic of this chapter.

Hatters, D.M. and Howlett, G.J., The structural basis for amyloid formation by plasma apolipoproteins: a review, *Eur. Biophys. J.*, 31, 2–8, 2002.

Higuchi, K. et al., Polymorphism of apolipoprotein A-II (apoA-II) among inbred strains of mice: relationship between the molecular type of apoA-II and mouse senile amyloidosis, *Biochem. J.*, 279 (pt. 2), 427–433, 1991.

˙Higuchi, K. et al., Fibrilization in mouse senile amyloidosis is fibril conformation-dependent, *Lab. Invest.*, 78, 1535–1542, 1998. (An interesting study showing that amyloid formation by ApoAII exhibits a phenomenon similar to prion strains.)

Higuchi, K. et al., Apolipoprotein A-II gene and development of amyloidosis and senescence in a congenic strain of mice carrying amyloidogenic ApoA-II, *Lab. Invest.*, 72, 75–82, 1995.

Iconomidou, V.A., Vriend, G., and Hamodrakas, S.J., Amyloids protect the silkmoth oocyte and embryo, *FEBS Lett.*, 479, 141–145, 2000.

Jarrett, J.T. and Lansbury, P.T., Jr., Seeding "one-dimensional crystallization" of amyloid: a pathogenic mechanism in Alzheimer's disease and scrapie? *Cell*, 73, 1055–1058, 1993.

Johan, K. et al., Acceleration of amyloid protein A amyloidosis by amyloid-like synthetic fibrils, *Proc. Natl. Acad. Sci. USA*, 95, 2558–2563, 1998.

˙Kane, M.D. et al., Evidence for seeding of beta-amyloid by intracerebral infusion of Alzheimer brain extracts in beta-amyloid precursor protein-transgenic mice, *J. Neurosci.*, 20, 3606–3611, 2000. (Reports the acceleration of Alzheimer's neuropathology in transgenic mice by seeding with amyloid plaques from a human brain with Alzheimer's disease.)

˙˙Kelly, J.W. and Balch, W.E., Amyloid as a natural product, *J. Cell Biol.*, 161, 461–462, 2003.

Kenney, J.M., Knight, D., Wise, M.J., and Vollrath, F., Amyloidogenic nature of spider silk, *Eur. J. Biochem.*, 269, 4159–4163, 2002.

Kimberlin, R.H., Experimental scrapie in the mouse: a review of an important model disease, *Sci. Prog.*, 63, 461–481, 1976.

˙King, C.Y. and Diaz-Avalos, R., Protein-only transmission of three yeast prion strains, *Nature*, 428, 319 323, 2004. (One of the best proofs for the protein-only nature of yeast prions.)

Kisilevsky, R. and Boudreau, L., Kinetics of amyloid deposition, I: the effects of amyloid-enhancing factor and splenectomy, *Lab. Invest.*, 48, 53–59, 1983.

Kisilevsky, R., Lemieux, L., Boudreau, L., Yang, D.S., and Fraser, P., New clothes for amyloid enhancing factor (AEF): silk as AEF, *Amyloid.*, 6, 98–106, 1999.

Li, G. et al., The natural silk spinning process: a nucleation-dependent aggregation mechanism? *Eur. J. Biochem.*, 268, 6600–6606, 2001.

Lindquist, S., Mad cows meet psi-chotic yeast: the expansion of the prion hypothesis, *Cell*, 89, 495–498, 1997.

Lomakin, A. et al., On the nucleation and growth of amyloid beta-protein fibrils: detection of nuclei and quantitation of rate constants, *Proc. Natl. Acad. Sci. USA*, 93, 1125–1129, 1996.

˙Lundmark, K. et al., Transmissibility of systemic amyloidosis by a prion-like mechanism, *Proc. Natl. Acad. Sci. USA*, 99, 6979–6984, 2002. (An important article showing evidence for a prionlike phenomenon in a systemic amyloidosis.)

Maddelein, M.L. et al., Amyloid aggregates of the HET-s prion protein are infectious, *Proc. Natl. Acad. Sci. USA*, 99, 7402–7407, 2002.

Masison, D.C. et al., [URE3] and [PSI] are prions of yeast and evidence for new fungal prions, *Curr. Issues Mol. Biol.*, 2, 51–59, 2000.

Price, D.L., Sisodia, S.S., and Borchelt, D.R., Genetic neurodegenerative diseases: the human illness and transgenic models, *Science*, 282, 1079–1083, 1998.

Rocken, C. and Shakespeare, A., Pathology, diagnosis and pathogenesis of AA amyloidosis, *Virchows Arch.*, 440, 111–122, 2002.

*Si, K., Lindquist, S. and Kandel, E.R., A neuronal isoform of the aplysia CPEB has prion-like properties, *Cell*, 115, 879–891, 2003. (A very interesting report showing evidence for a new prion in aplysia involved in memory formation.)

**Sigurdsson, E.M., Wisniewski, T., and Frangione, B., Infectivity of amyloid diseases, *Trends Mol. Med.*, 8, 411–413, 2002.

Sipe, J.D. et al., Human serum amyloid A (SAA): biosynthesis and postsynthetic processing of preSAA and structural variants defined by complementary DNA, *Biochemistry*, 24, 2931–2936, 1985.

Skinner, M., Shirahama, T., Benson, M.D., and Cohen, A.S., Murine amyloid protein AA in casein-induced experimental amyloidosis, *Lab. Invest.*, 36, 420–427, 1977.

Soto, C., Protein misfolding and disease: protein refolding and therapy, *FEBS Lett.*, 498, 204–207, 2001.

Sparrer, H.E., Santoso, A., Szoka, F.C., Jr., and Weissman, J.S., Evidence for the prion hypothesis: induction of the yeast [PSI+] factor by in vitro-converted Sup35 protein, *Science*, 289, 595–599, 2000.

**Stefani, M. and Dobson, C.M., Protein aggregation and aggregate toxicity: new insights into protein folding, misfolding diseases and biological evolution, *J. Mol. Med.*, 81, 678–699, 2003.

*Tanaka, M., Chien, P., Naber, N., Cooke, R., and Weissman, J.S., Conformational variations in an infectious protein determine prion strain differences, *Nature*, 428, 323–328, 2004. (One of the best proofs for the protein-only nature of yeast prions.)

Tanaka, M., Chien, P., Yonekura, K., and Weissman, J.S., Mechanism of cross-species prion transmission: an infectious conformation compatible with two highly divergent yeast prion proteins, *Cell*, 121, 49–62, 2005.

Telling, G.C. et al., Transmission of Creutzfeldt-Jakob disease from humans to transgenic mice expressing chimeric human-mouse prion protein, *Proc. Natl. Acad. Sci. USA*, 91, 9936–9940, 1994.

**Uptain, S.M. and Lindquist, S., Prions as protein-based genetic elements, *Annu. Rev. Microbiol.*, 56, 703–741, 2002.

Walker, L.C. et al., Exogenous induction of cerebral beta-amyloidosis in betaAPP-transgenic mice, *Peptides*, 23, 1241–1247, 2002.

*Wickner, R.B., [URE3] as an altered URE2 protein: evidence for a prion analog in *Saccharomyces cerevisiae*, *Science*, 264, 566–569, 1994. (A seminal paper that expanded the prion concept beyond mammalian prions.)

Wickner, R.B. et al., Prions of yeast and fungi: proteins as genetic material, *J. Biol. Chem.*, 274, 555–558, 1999.

**Wickner, R.B. et al., Prion genetics: new rules for a new kind of gene, *Annu. Rev. Genet.*, 38, 681–707, 2004.

Wisniewski, H.M., Merz, G.S., and Carp, R.I., Senile dementia of the Alzheimer type: possibility of infectious etiology in genetically susceptible individuals, *Acta Neurol. Scand. Suppl.*, 99, 91 97, 1984

Wosten, H.A. and de Vocht, M.L., Hydrophobins, the fungal coat unravelled, *Biochim. Biophys. Acta*, 1469, 79–86, 2000.

*Xing, Y. et al., Transmission of mouse senile amyloidosis, *Lab. Invest.*, 81, 493–499, 2001. (An important report providing compelling evidence for the transmissibility of apoAII amyloidosis.)

Xing, Y. et al., Induction of protein conformational change in mouse senile amyloidosis, *J. Biol. Chem.*, 277, 33164–33169, 2002.

Index